THE CONTINUED EXE

THE CONTINUED EXERCISE OF REASON

PUBLIC ADDRESSES BY GEORGE BOOLE

EDITED AND WITH AN INTRODUCTION BY
BRENDAN DOOLEY

The MIT Press
Cambridge, Massachusetts
London, England

This book was set in Bembo Std by Toppan Best-set Premedia Limited. Printed and bound in the United States of America.

Library of Congress Cataloging-in-Publication Data

Names: Boole, George, 1815-1864, author. | Dooley, Brendan Maurice, 1953- editor.
Title: The continued exercise of reason : public addresses by George Boole / edited and with an introduction by Brendan Dooley.
Description: Cambridge, MA : The MIT Press, [2018] | Includes bibliographical references and index.
Identifiers: LCCN 2017032612 | ISBN 9780262535007 (pbk. : alk. paper)
Subjects: LCSH: Boole, George, 1815-1864. | Logic, Symbolic and mathematical.
Classification: LCC QA9.2 .B657 2018 | DDC 511.3--dc23 LC record available at https://lccn.loc.gov/2017032612

10 9 8 7 6 5 4 3 2 1

CONTENTS

CONTENTS

Alexander Wheelock Thayer introduces an exhaustive life of Beethoven by remarking, "There is but one road to excellence, even for the genius of a Handel or a Mozart—unremitted application."[1] Yet where does this road pass? Through a place? A time? Trying to understand a great thinker from long ago involves the researcher in a curious conundrum. We seem to know the conclusion—the published work, the famous insights—but what are the premises? George Boole received the Royal Society's first-ever gold medal in mathematics in the year 1844, and his crowning achievement, *The Laws of Thought*, published ten years later, stands as a monument of mathematical thinking. What clues are left about how those accomplishments occurred? We have a myriad of private papers scattered here and there between Britain and Ireland. If possession of an exceptional mind is not enough, what were the necessary or sufficient conditions that made the work a possibility?

In advance of a comprehensive biography that someone may someday write, this book introduces the figure of George Boole by perhaps the next-easiest approach: through his own words. Bicentenary celebrations such as in the recent "Boole Year" (2015) often seem to allow the present to overwhelm the past. Publishing these texts will hopefully steer some of the interest back to the man himself. In fact, the closer we look, the more material his writings seem to offer for reflection by a wide variety of readers, including those whose chief interests may lie in the histories of science, culture, religions, education, and even freedom. Partly due to the nature of the occasions for which these texts were written, mathematics—Boole's recognized area of competence—takes a back seat here, yet the reader will find that everything in one way or another relates to the laws of thought.

The Boole we are coming to know was not simply a polymath but also an intellect of surprising originality in numerous fields of knowledge—indeed, an intellect keenly interested in how those fields fit together, and how they might contribute to the grand project of human engagement with the real and the ideal, the sacred and the profane, the good and the contrary. Investigators into any question, in his view, must attend to the operations of logical abstraction that characterized all thought processes. Everywhere patterns abound—of behavior, expression, belief, experience, memory—which must be sought out and made to reveal the structures underneath, even the laws, applying not only to the bodies in the universe and the things on earth but the human story too. In the effort to reach a deeper understanding, Boole's answers are sometimes surprising, sometimes prescient, sometimes disconcerting; they are illustrated in the pages below.

Although this collection does not refer to the history of mathematics per se, the editor was drawn to the subject by curiosity about the tension between what appear to be two poles of methodological reflection: science and humanities. In the early nineteenth century, that opposition was far less clearly drawn than now, although at the time of this writing there are conciliatory movements afoot on both sides of the "two cultures" divide famously delineated by C. P. Snow in 1959, and not only because of the twilight of positivism. Themes of methodological cross-fertilization have come up in the work of Robert K. Merton and a host of others, but perhaps nowhere quite so explicitly as in Edward O. Wilson's *Consilience* (1998), which argues that knowledge can be united by a "continuous skein of cause-and-effect explanation and across levels of increasingly complex organization"—a concept robust enough to serve as the framework for a conference held at the New York Academy of Sciences (proceedings published in the academy's *Annals* in 2001), which concluded with a hope for someday realizing an ideal that harks back to humanity's basic desire to know.[2] As the debate continues, and the seemingly weaker sciences compete against the supposedly stronger, while the notion of "truth" gives way to "truth communities," Boole's work recalls the universality of reason.

On another personal note, at Jacobs University, located in Bremen, Germany, a history professor taught a course with a professor of mathematics under the title the Mathematics of History, ranging from the ancient world to modern times, including Archimedes, Leonhard Euler, Niels Henrik Abel, Évariste Galois, and Albert Einstein, with excursions into Charles Babbage, Boole, and many others. Joint lectures would divide the material about each figure into contents and contexts. The students were hardworking, and the discussions with my colleague Ivan Penkov always highly stimulating. A major assumption of the course was that the two disciplines, math and history, had something in common in the realm of fact and proof, at least to the extent that organized thoughts in a logical sequence led ineluctably to our conclusions, because numbers and deeds were creatures of the human mind. Maybe I am a little less certain about this analogy now than I was then, but there seems little doubt that Boole was as much a creature of mathematics as he was of history. This book is devoted to helping find out how.

Great minds may occasionally appear somewhat one-dimensional, at least in the way they are portrayed. George Boole, for instance, was a pioneer of the information age for more reasons than the one usually cited: that he developed the system of algebraic logic which eventually found an unexpected engineering application in the design of switching circuits.[1] He was also an early advocate of the mass distribution of knowledge, using the methods at his disposal in early Victorian times. His range was impressive, his curiosity immense. In the classroom and lecture hall, he interpreted the results of recent discoveries and debates originating among specialists in numerous fields—history, psychology, ethnography, and much else—and communicated them to a broad audience. At a time when the first steam-operated printing presses were churning out masses of material of uneven quality on every possible subject, he helped audiences to make sense of what they heard and read. Less known and therefore less appreciated is Boole's role in the history of the making and organization of knowledge. A better understanding of this feature, perhaps inspired by the works on display here, may eventually provoke a more thoroughgoing reappraisal of the whole figure.

Within the histories of mathematics and philosophy, Boole's position is well established; indeed, his contribution to symbolic logic can be summed up in a phrase: symbolic logic. Building on work by George Peacock and a strand of research going back to Gottfried Wilhelm Leibniz, he extended the traditional syllogistic method inherited from Aristotle by developing a systematic way of symbolically representing logical propositions using algebraic notation that added clarity and precision to the processes of logical generalization and abstraction.[2] In this way, arguments could have more premises and involve more classes than before, so

Boole provided theorems allowing such complex reasoning to be handled by algorithms. The major breakthrough was soon generally recognized, although not always assigned to his name, and by the end of the nineteenth century, Charles Lutwidge Dodgson (better known as Lewis Carroll, whose literary pursuits have nowadays eclipsed the mathematical ones) devised a program for teaching symbolic logic to children.[3]

The contribution did not stop there. In 1901, Bertrand Russell strikingly affirmed that "pure mathematics was discovered by Boole." In his view, *The Laws of Thought* was nothing short of "the first [book] ever written on mathematics." Of course, such a position depended on the definition. For Russell, as for modern practitioners, mathematics is a system for understanding modes of analysis, methods of problem solving, and relations between mathematical objects—as distinct from the actual results obtained by particular operations, or by application to practical problems in, say, physics and engineering. The science of abstraction and generalization exists (according to this view) in a realm apart from, yet somehow related to, the physical world—especially to the extent that any question can be expressed in terms of symbols subject to a set of rules, which is the essence of the discipline. "Pure mathematics," Russell goes on to say, "consists entirely of assertions to the effect that, if such and such a proposition is true of anything, then such and such another proposition is true of that thing." Indeed, "it is essential not to discuss whether the first proposition is really true, and not to mention what the anything is, of which it is supposed to be true."[4] Until Boole emancipated pure from applied mathematics, the two were viewed as one and indivisible. But such innovations were only part of Boole's activity as a cultural figure.

Before Boole joined a university faculty as a professor of mathematics in 1849, public science and education occupied much of his time. In the unique cultural environment of 1830s and 1840s Britain, social improvement and cooperation were the watchwords of the day, and he was deeply committed. "The continued exercise of reason," he opined, "was a tonic for the brain, rivaled only by practical engagement of the eye and the hand with the materials from which reasonable conclusions were to be drawn.[5] In his speeches and writings, he drew on a host of influences from current intellectual trends that he either mentioned or implied: scientific debates, rumors, and media sensations. Bringing this context back into focus may add clarity to an emerging picture of his significance as a public lecturer and of his possible impact in a key moment of transformation in society, economics, and politics. Seen from another perspective, his activity throws more light on a time and place that are often viewed in terms of ambiguous leanings, deeply rooted in the past but seemingly directed toward the future.

For gaining an appreciation of Boole's interests and methods as a thinker and educator on a wide variety of topics, few actual lectures remain, and those few are occasionally cited but almost never read. Dating from his years in Lincoln, England, "On the Character and Origin of the Ancient Mythologies," "On the Question: Are the Planets Inhabited?," "A Plea for Freedom," and "On Education," published here for the first time, as well as "On the Genius and Discoveries of Sir Isaac Newton" and "The Right Use of Leisure," printed for the first time in 180 years, plus "The Claims of Science," which last appeared in a rare edition in 1952, and "The Social Aspect of Intellectual Culture," reprinted for the first time since 1855, offer precious insight into the early thinking of the man who became the first mathematics professor at Queens College in Cork (later UCC). Delivered in various venues, they also reveal the ways and means of popular education in the age of John Stuart Mill, a great contemporary with whom Boole shared not only an interest in logical reasoning but also in social improvement.

Born into the family of a cobbler whose cultural aspirations and knack for intellectual pursuits inspired young George to choose books over trade from early on, by age fourteen he was a published poet. At sixteen he assumed financial responsibility for his mother and younger siblings. He thus took up a teaching position in Doncaster, not far from home— the first of a number of such assignments carried out with increasing success and confidence.[6] By age nineteen, he established his own school in Lincoln and soon took over as headmaster of another, older one. That was also the year he addressed the Lincoln Mechanics' Institute on the topic of Isaac Newton, of whom the institute had recently acquired a portrait bust for its meeting room, and spoke to such good effect that his lecture was designated for immediate publication.[7] Over the years in Lincoln and environs, he became known equally as an educator and lecturer to the wider public on a variety of topics generally handled with especial attention to the rational steps of each argument and the empirical grounding.

"It is one of the distinguishing characteristics of our human nature," Boole said in one of his lectures, "that we are enabled to penetrate beyond particular facts and to discern the general principles which underlie them."[8] To realize this and the other great potentialities of the mind, however, required adequate training. Boole himself received little formal schooling, but he benefited from the casual tuition of a variety of adults in his midst as well as meetings of local associations devoted to adult education. And throughout his career, he accompanied his scientific endeavors with a commitment to communicating what he knew to a wider audience. As an educator, he exemplified the historical transition from exclusively text-based to experiential approaches to learning, aimed at engaging not only elite but also popular audiences.[9] The documents regarding his activities reveal the startling newness of an age when yearnings awakened by opportunities for advancement flattered social ambitions, bearing eloquent witness to the power of knowledge in early Victorian Britain.

There was much to improve, many thought, even in an age that had already ushered in the railway, steam press, telegraph, and photography. As the Industrial Revolution spawned new sources of wealth and emerging technologies suggested new ways for humanity to cope with the world, old questions began to resurface with new urgency. If progress was inevitable, as philosophers and historians agreed, why were poverty and inequality so deeply rooted? If Britain was a beacon of civilization to the world, as colonial propagandists proclaimed, why did misery seem never less? If current arrangements guaranteed justice to all, as advocates promised, why so much discontent? The more concentrated intelligence could be deployed to answer such questions, so the proponents of public instruction believed, the better off humanity would be.[10]

Hopes and fears seemed set on a collision course. A widespread population boom, according to David Ricardo, supplied more ready hands for work. Yet followers of Thomas Malthus warned about the dangerous limits on the distribution of necessities under the prevailing regime. Political and economic thought seemed more intent on conserving the status quo, while the few prophets of doom suggested worrying scenarios where the winners of today might be the losers of tomorrow. The specter of communism soon to be announced by Karl Marx was not yet stalking the earth. But no one forgot the 1819 Manchester riots of hungry crowds inspired by radical leaders, brutally put down. Socialists like Robert Owen, convinced that things could not stand as they were, actually carried their utopian programs out in practice between Great Britain and the United States. Mill, attempting to characterize "the spirit of the age," said, "The first of the leading peculiarities ... is, that it is an age of transition. Mankind have outgrown old institutions and old doctrines, and have not yet acquired new ones."[11]

While the human sciences inherited from the Enlightenment seemed mostly inadequate to take on the tasks ahead, emerging political economy attempted to address the present predicament systematically, but there was much debate regarding the mode and method.[12] In suggesting solutions to present problems, should laws and principles prevail, as the Enlightenment philosophers had proposed and the utilitarians suggested? Or should present and past experience be sifted, scrutinized, compared, and made to reveal underlying patterns that might serve as guides to action? Deduction or induction, or both, and in what measure? Dissatisfaction with the presumed limitations of long-running organizations such as the London Royal Society and its counterpart in Edinburgh led Charles Babbage and Adam Sedgwick in 1831 to form the British Association for the Advancement of Science, embodying a definition of useful studies that included the tripartite remit of humans, nature, and society. An affiliated meeting in Cambridge including Adolphe Quetelet, a pioneering social statistician and sociologist, along with William Whewell, Richard Jones, and Malthus, confirmed the methodological rigor of social science. Yet an authoritative view divided theory and practice, insisting that any examination of current structures should avoid "communication with the dreary wild of politics."[13]

Meanwhile, the dreary living conditions of masses of people were being turned into immortal literature by Charles Dickens, whom the *Edinburgh Review* proclaimed as "the most popular writer of his day," and whose striking descriptions of the urban underclass, conveyed with style and wit, inspired readers' compassion for characters like Oliver Twist, the parish boy whose tortuous "progress" made snide allusion to the pilgrim's progress in John Bunyan's Christian classic. Instead, in Dickens's version, miserable circumstances, the seedbed of immorality and vice, challenged the virtue of even the most innocent. Any troubled reader tempted by the happy ending to think that all's well that ended well only had to recall the

seemingly infinite trail of unfortunate individuals driven to navigate piti-
fully throughout these pages between dignity and death in a diabolical
system. "He especially directs our attention to the helpless victims of
untoward circumstances," the reviewer added, and on the whole, "the
tendency of his writings is to make us practically benevolent—to excite
our sympathy in behalf of the aggrieved and suffering in all classes; and
especially in those who are most removed from observation."[14]

Lincoln was no London, to be sure, but the town's population more
than doubled between 1800 and 1840, bursting through its medieval limits.[15] The whole county grew by nearly 75 percent, in tandem with the
East Midlands manufacturing area, and was not spared by Friedrich
Engels in his pathbreaking work *On the Condition of the Working Class in
England*, published in 1845. This condition "is the real basis and point of
departure for all social movements of the present," Engels asserted,
"because it is the highest and most unconcealed pinnacle of the social
misery existing in our day." The formation of a landless peasantry, working fields for inadequate pay on great estates now run for money, not
noblesse oblige, he pointed out, went accompanied by "social war." A
mythical "Captain Swing" supposedly waged a campaign of farm burnings, actually done by disgruntled laborers. According to contemporary
testimonies, among the many other such burnings elsewhere, no fewer
than twelve occurred in Derby, Lincoln, and the South—not including
the deliberate destruction of farm machinery. The "quiet, idyllic country
districts of England," Engels proclaimed, were fast losing their charm, and
current research, at least along general lines, vindicates this view.[16]

The world was evidently on the verge of a "great transformation," but
the outcome was still impossible to imagine. Nor has that phrase, coined
long ago by Karl Polanyi, lost any of its poignancy, although the underlying features have changed character with the advances in research. To be
sure, Georg Wilhelm Friedrich Hegel already recognized a pivotal epoch
in political economy in 1821, when he remarked that developments
could not continue unchecked. "Control is ... necessary to diminish the
danger of upheavals arising from clashing interests and to abbreviate the
period in which their tension should be eased through the working of a
necessity of which they themselves know nothing."[17] The British liberal
elite and their allies in the worlds of business and finance (among others)
looked on in horror as France seemed to lurch from rebellion to

revolution and back again with no end in sight.[18] They came to believe that the massive destruction of lives and property at home could only be averted by carefully measured doses of electoral reform.[19] After the defeat of the Conservative Wellington ministry and the advent of a government led by the Whig reformer Charles, second Earl Grey, passage of the limited suffrage law of 1832 initiated a process that would result also in the New Poor Law of 1834, designed to stop handouts to the poor, sweep the hard cases off the streets, and put the able bodies to work in workhouses where they might at least learn a trade.[20]

Not good enough, said the Chartists, agitating for more radical change with the rallying cry "universal suffrage or universal revenge."[21] Their proposed bill of 1838 demanded a new generation of worker-friendly parliamentarians, to be brought in by abolishing property requirements for membership and providing a subsistence allowance during session. Sporadic demonstrations around the country encountered police repression, jailings, injuries, and deaths. When presented to Parliament for the final time in 1848, the unwieldy petition with its more than three million signatures (not all authentic, claimed the critics) had to be brought in on three carriages—to no effect, at least for the moment. Giving the vote to the masses, as now constituted, was unthinkable according to those best able to assert their opposition by money and force.

Whenever an effective consensual government might occur, education had to come first. And around the doctrine of self-improvement through individual instruction and work, the elite made common cause with a number of the popular leaders. Meanwhile, education became a booming business in Britain, especially in manufacturing areas like Lincolnshire, where the population increase was accompanied, after much agitation, by the Factory Acts.[22] Children ages nine through thirteen could now work no more than nine hours per day (at least not legally), and children above that age, no more than twelve. What would they do with their spare time? The 1833 bill mandated at least two hours of instruction per day for the youngest. According to the moralists of the period, here was an opportunity, however modest, to put program into practice. No one ignored the limitations. Against the notion of lowering adult hours, the laissez-faire doctrine embraced the freedom-loving argument that no one should be forced to reduce their workload to ten hours or less against their will.[23] But there were plenty of adults in small trades and businesses with some

leisure moments available for mental exercise and culture in the broadest
sense—or such was the hope.

Leisure time was also the particular concern of a group called the
Early Closing Association, founded in London around 1843 and diffused
to Lincoln shortly thereafter, aimed at encouraging owners to reduce
the long hours worked by tradespeople in shops and other small businesses in the retail sector not covered by the Factory Acts. Combining a
social agenda with a religious one, the group advocated leisure not only
for the educational goals of self-improvement but also for religious
activity seemingly precluded by the prevalence of profane occupations
on the Sabbath. Boole professed full agreement with a combined social
and spiritual agenda without referring, at least in this context, to organized religion. As vice president of the Lincoln branch of the association,
he spoke about "The Right Use of Leisure" in a lecture printed here,
exhorting his audience in these words: "Would that some part of the
youthful energy of this present assembly might thus expend itself in
labours of benevolence! Would that we could all feel the deep weight
and truth of the divine sentiment, that 'No man liveth to himself, and no
man dieth to himself.'"

Cooperation and community expressed in terms of brotherly love: the
formula was a powerful motivator. For many in the roles of thought and
learning, intellectual activity acquired a new raison d'être.[24] Joy, and even
some glory, might derive from communicating serious ideas to a roomful
of eager listeners from a variety of backgrounds. Beyond the fleeting
pleasure of a brief applause, there was the enduring satisfaction of contributing to the long-term benefits of peace, prosperity, and self-
improvement heralded by moralists and reformers, and proclaimed widely
in the press. Boole had only to think about the role public lecturing had
played in his own education to be convinced about the role it might play
in the lives of others. Would they be so profoundly influenced as he was?
All he and his fellow lecturers could do was wish and wait.

To what degree Boole consciously shared in some of the deeper agen-
das of the public education movement can only be conjectured. Inevitably, for some, control was more urgent than emancipation. The wonders
of human beings and nature should induce piety not arrogance, obedience not presumption. Leonard Horner, in the *First Report of the Directors
of the School of Arts* (1822), argued that among other blessings, a course of

lectures might advance the cause of religion by inspiring respect for "those beautiful contrivances, by which the Almighty has adapted the whole system of the universe to the comfort and advantage of man, and which at once display the infinite wisdom and goodness of an all-perfect Being."[25] Often expressed were the needs to redirect worker energies away from destructive and disorderly conduct, and toward behavior tending to conserve and advance present society and institutions. According to many of the leaders, workers engaged in the quiet pursuits of reading, listening, and conversing about the liberal arts were not, presumably, engaging in mass demonstrations, forming seditious conventicles, or behaving dangerously in any way. Thus were the long-term interests of all classes most admirably served.

The Lincoln branch of the London-based organization gained early recognition for significant accomplishment, and apparently by 1834 already counted some five hundred members. A particularly prescient move, in the opinion of Frederic Hill, reporting on the results of surveys taken in the various locations where Mechanics' Institutes took root, had been the engagement of a female librarian, whose presence on the premises had encouraged women to attend meetings—a precious datum for the larger drama of women's emancipation just then unfolding within British society as a whole.[26] All in all, effused a highly satisfied George Birkbeck, founder of the movement, the Lincoln branch had "concentrated more influence, and more general power than had ever attended one institution," and hosted lectures "on all branches of moral philosophy." He did not have to add that the proscription of discussions about religion and politics seemed wholly appropriate not only for avoiding sectarianism but also for diverting attention from the anxious questions of franchise that otherwise agitated the public sphere. Now was the time for joining it formally to the London branch, Henry Pelham added, considering the informal and personal ties already in place, and also because his father, Charles Anderson-Pelham, Baron Yarborough (and from 1837, first Earl), former member of Parliament for Lincolnshire, happened to be its protector.[27]

Audience impact is notoriously difficult to measure, especially at a distance of nearly two centuries. Typically the researcher possesses somewhat adequate information on the production side: announcements listing lecture titles and perhaps, as in the present case, texts that must closely resemble what was actually heard. Boole's activities were well enough appreciated by the leadership in his organization to gain him multiple places on the roster and eventually an invitation to take over as president, which he refused due to other commitments. Reports in the local newspaper, the *Lincolnshire Chronicle*, typically gave testimony to, for instance, the "sound learning and great research" evidenced in particular lectures, such as "On the Question: Are the Planets Inhabited?"[28] Any reference to audience appreciation was couched in the most generic terms, such as "the room was well filled, and the lecture, which occupied an hour and a half in the delivery"—in this case, "The Right Use of Leisure"—was "listened to with the attention it deserved."[29]

That the Mechanics' Institutes and their affiliates, at least in the first decades, may have attracted a public mainly among the middling ranks rather than, as originally intended, the "mechanics"—that is, skilled workers (according to the somewhat misleading jargon of the time)—seems to have made little difference in regard to their overall significance.[30] Their message apparently appealed to many, as confirmed by their more or less flourishing condition throughout the troubled 1830s and 1840s, and well into the 1850s. Martyn Turner, arguing against a prevailing historiography, suggests that the first two decades were only the prelude to an ever more incisive involvement in local communities.[31] On this reading, Boole's activities helped set the stage in an influential preparatory phase in the history of public education, which eventually received new impetus locally from the Great Exhibition and abroad from increasing competition by industrialized countries. With the Education Act of 1870 and subsequent legislation tending to provide widespread free instruction, a major objective of the movement arguably had been finally achieved—just as the season of European liberalism was drawing to a close.

Whatever may have been the intentions of those involved, abundant evidence points to a new orientation in society and politics. Already in Boole's time, public discussion in British society was a formidable force, now operating on numerous channels, in public squares, tearooms, public houses, auditoriums, meeting halls, and print. Notably, several of his discourses were later published. Apart from the significance of the specific notions and topics, whether about science, society, religion, history, politics, or whatever else, trends already developing in the previous century gained new impetus. Below the surface of discussion the structures of an alternative power in society were being formed, distinct from the institutions of public administration. An influential thesis, modified in the course of historiographical debates but not rejected, traces the occurrence of something new in the world: a public sphere of individuals making public use of their reason. On this view, what began in Britain and France already in the course of the eighteenth century would commence later in Germany and other countries, and there was no turning back. The creation of these structures would form a good part of the world in which we live.[32]

In the midst of his public activities, Boole developed a formidable research program out of the studies on which he had long been embarked.[33] The correspondence with experts like Duncan F. Gregory in Cambridge and Augustus De Morgan in London helped him stay abreast of new developments in spite of distractions due to job and family.[34] There were numerous papers in the *Cambridge Mathematical Journal*, and his "On the General Method in Analysis" for the *Philosophical Transactions of the Royal Society* (1844) received a special prize.[35] In due course a major book followed, *The Mathematical Analysis of Logic* (1847), and the world of learning was impressed. In this prelude to his later book, *The Laws of Thought*, Boole laid the groundwork for joining two hitherto-unrelated fields. Mathematical analysis so far had largely been conceived as an aspect of the "expression of magnitude" including "operations upon magnitude," wherein "the elements to be determined" are "measureable by comparison with some fixed standard." Even Newton, inventor of calculus some 150 years before, seemed to rely on the form-filled world of geometry for illustrating his propositions. On the other hand, logic was still regarded as a branch of philosophy—one of the three pillars of the academic trivium along with grammar and rhetoric, with which it shared an interest in language and the types of arguments used in every kind of discourse. Joining mathematical analysis to logic through the realization that both benefited from "the employment of symbols, whose laws and combinations are known," meant nothing less than the establishment of "a philosophical language" of great potential value "as an instrument of scientific investigation."[36]

The time seemed ripe for his educational work, like his mathematics, to acquire a more adequate institutional setting—but which one? As an outsider to the university system, his choices were limited. On hearing of professorships available at the newly founded Queen's College in Cork, and encouraged by friends, he rushed to apply. Concerning the prospect, he wrote to his informant William Thomson, "I should certainly prefer an appointment of the kind you name." Compared to a position in "the mastership of a private school," he explained, "it would be free from that uncertainty and dependence which are inseparable from the profession of a schoolmaster at least in the present state of society." Finally, "as respects the prosecution of mathematical studies I might look for advantages also which I do not at present enjoy."[37]

The decision to move was a leap in the dark, made riskier by the emerging circumstances. Already at the end of 1846 the protracted decision-making process within Irish officialdom concerning his appointment confirmed his suspicions. "I hear nothing from Ireland," he commented, guessing the reason: "The wretched state of the country physically and morally I suppose quite enough to tax the energies of any government."[38] News just then began trickling in about the extent of the most recent mortality following a nearly complete crop failure, compounding the devastations of the previous year provoked by the onset of the potato blight. There was unrest, and in Cork, methods tried and tested in the Chartist riots in England were put to good effect. Ringleaders were rounded up, and in one case, a youth was shot.[39] As the disaster spread, inadequate policies based on misinformation and prejudice only made things worse, and the story told many times needs no retelling here.[40] Ireland, for the time being, was doomed, but Boole stuck to his resolve.

Another English writer, Anthony Trollope, lived about a day's ride from Cork while scouting the country on troubleshooting missions for the Irish Post Office. In what he later recalled as an altogether "jolly time," he spent the wee hours writing each morning and produced two books set in Ireland: *The Macdermots* followed by *The Kellys and the O'Kellys*. They were inspired, he later said, by a place that was gone. During the last stages of publication, that place was erased. "In 1847 and 1848 there had come upon Ireland the desolation and destruction, first of the famine and then of the pestilence which succeeded the famine. It was my duty at that time to be travelling constantly in those parts of Ireland in which the misery and troubles thence arising were, perhaps, at their worst. The western parts of Cork, Kerry and Clare were pre-eminently unfortunate."[41] Although he remained in the country until 1859, there would be no more novels set there; only a series of letters to the *Times* gave voice to his impression about the way things were.

Almost as widespread as the reports on Ireland's plight were the suggestions for change. Robert Kane, an accomplished chemist trained at Trinity College in Dublin, and from 1845 the first president of Queen's College in Cork, focused on the long term. Policies must be aimed not only at developing the considerable natural resources of the country, he insisted, but also at developing human resources through improved education for the challenges of an increasingly industrialized world. "The education necessary for industrial pursuits is very generally underrated in this country, and from this cause alone springs a great deal of our want of industrial knowledge," he wrote in a four-hundred-plus-page treatise, *The Industrial Resources of Ireland*, published on the eve of the famine. Too often, he insisted, skills in the practical arts relevant for the invention and manufacture of merchandise were left to chance and on-the-job experience—unlike the skills necessary in professions such as law, medicine, or theology, rigorously defined by curricula in schools and universities. And yet "every department of agricultural and of manufacturing industry has its origin in scientific principles, practically applied." As examples of areas where the "practical sciences" contributed insights for productivity and job creation, he cited "the weaving of a woolen cloth, the rolling of an iron rail, or of a brass ornament, the construction of a clock, the preparation of soap, or of oil of vitriol," all of which "require the discovery of a certain principle, which, working by a certain process or certain materials, elaborates the product."[42] Semi-official lectureships on such matters, such as the Liberal government had been encouraging around the country, were, in his view, a good start. Indeed, he had lent his own services in this regard, along with other talented and popular demonstrators of the more spectacular aspects of natural knowledge applied to technology, such as William Lover, an audience favorite in the areas of magnetism, electricity, and chemistry. But a fundamental shift in educational institutions had to take place for real progress to occur.[43] The new Queen's Colleges (in Cork, Belfast, and Galway) thus had a basic role to play in rebuilding the country, and the going would not be easy.

Whatever reasons may have combined to make Boole accept an offer in such a troubled country far from home, he took on the new challenge with his customary aplomb. Soon after the college's doors opened in fall 1849, he was in front of a classroom teaching mathematics to some fifty undergraduates per year, reportedly to good effect. The sharpest memories regarding his performances tend to be of exceptional occasions, such as when the distracted professor, writing at the blackboard with his back to the benches before the bell rang, failed to notice students file in, sit down for a time, and file out, but turning around finally and seeing no one, concluded that the class had been canceled and headed home.[44] Posthumous tributes bear witness to a painstaking instructional approach appreciated for the attempt to accommodate unequal abilities. "Great at the black-board," reported one. Said another: "He appeared to lower himself to our standard for the time being, and having thus gained our sympathies he rapidly rose and carried us along with him."[45] Any deeper insights into what transpired in the Irish classroom, apart from the official course announcements, will have to await a fuller analysis of the few existing lecture notes, which have so far resisted interpretation.

Although an individual of Boole's versatility must have been a particular asset for the young institution, his impact on his surroundings is only now being studied after years of comparative neglect. Outside the classroom, he continued his mission to instruct the general public on key issues of perennial importance for which the opinions of a dedicated intellectual might hold some weight. Here I include two public addresses from the Cork years: one titled "The Claims of Science," delivered at the opening of the Queen's College academic year to an audience including university patrons and well-wishers along with students and faculty from every department, and another titled "The Social Aspect of Intellectual Culture," penned for the opening of a small exhibition by the Cuvierian Society featuring items of intellectual, cultural, and commercial interest. In addition to the positive and uplifting message in both speeches, realistic according to the standards of the day, and framed within the theological and devotional context that we have come to expect, he attempted to address the critical problem of the organization of knowledge and structure of the disciplines, precisely where his research was breaking new ground.

The material from both sides of the Irish Sea forms a corpus of remarkable pith and vivacity, testifying to a distinguished career in public speaking. And the figure of Boole that emerges from these pages is of an exceptionally gifted man who also belonged to his own time, and whose life casts much light on an era of great minds grappling with the unsolved conundrums of existence. Recirculating his ideas on a wide variety of topics may hopefully contribute some sharp insights well worth further exploration, while encouraging a closer look at a key episode in European cultural history that left a profound impact on a shared modernity that defines much of the global society impinging on all of us. As an aid to comprehension as well as enjoyment, I now embark on a more detailed account of the particular characteristics and contributions of each lecture, along with the precise occasion, when known.

ON THE GENIUS AND DISCOVERIES OF SIR ISAAC NEWTON

Boole's first-known public address celebrated the unveiling of a statue of Newton donated by the Duke of Yarborough to the reading room of the Lincoln Mechanics' Institute in 1835. For a young man of nineteen, it was chance to express something distinctive while appealing to an audience he respected. In view of the obvious expectations in this venue, he chose as his theme an examination of the roots of Newton's practical ingenuity, in light of the accumulated wisdom of the subsequent century and a half. The result was so well appreciated that an order went out for immediate printing, and that is the form in which I present it here.

The choice of subject was dictated by the occasion, but evoking the figure of Newton was timely in any case. A kind of personality cult had grown up around the pioneering natural philosopher, due in no small measure to his real accomplishments, enhanced to a considerable degree by myth and legend; indeed, just across the English Channel, Claude Henri de Rouvroy, comte de Saint-Simon, had developed a social theory around a Newton-based "religion."[46] The emerging profession of science needed heroes, and here was one ready-made. No matter that a significant portion of the work was in fields like theology and alchemy, where the conclusions were no longer of much interest to a public fascinated by demonstrations in physics and dynamics.[47] The figure of a new

Archimedes seemed to appeal to a growing vogue for the benefits of practical technology that were rapidly changing the daily practices of modern life.

A dogma of the Mechanics' Institutes associated the name of Newton with that of James Watt, considering that the invention of the steam engine seemed to have ushered in a mechanical revolution at least as profound as the intellectual revolutions conceived by the philosophers. Such was the theme developed by Thomas Hodgskin, a journalist, evoking both names in an address to the London Mechanics' Institute in 1827. Whether Boole shared the same sentiments is hard to say. Perhaps by background, he would have had more in common with the Glasgow instrument maker (Watt) than with the Cambridge don (Newton). But what Hodgskin said of both applied equally well to him: "master-spirits who gather and concentrate within themselves some great but scattered truths, the consequences of numberless previous discoveries."[48]

With locks of Newton's hair going for three times as much as the martyred King Charles I's on the memorial relics market of the second decade of the century, and a purported Newtonian tooth set in a gold ring having fetched some £730, there is no wonder that sculpted likenesses of the scientist par excellence appeared in innumerable places around Britain.[49] No matter that the man's actual visage was imperfectly recorded in the premortem iconography. Whatever he was thought to have looked like was mass-produced in busts and plaques even by the Wedgwood pottery works for display in homes and offices. And by Boole's day, laboratories, reading rooms, philosophical societies, and benevolent associations eagerly sought an instant gravitas ennobled by the authoritative gaze of the illustrious forebear.

Quite likely, Boole felt a certain kinship with Newton even before embarking on the lecture. Both, after all, were Lincolnshire men, and both demonstrated signs of prodigious mathematics ability at an early age. Was he thinking of his own case when he suggested, "In the intellectual, as in the physical constitution, a too early development is often followed by a feeble and transient maturity"? And was there a hopeful self-reflection in the observation, "The youthful occupations of Newton rather manifest the workings of an ardent and inquiring spirit, than of a mind already invested with the features of genius"? Other possible correspondences are harder to verify. Did he feel a religious affinity—for instance, with Newton's anti-Trinitarianism? At least according to the testimony of one who knew Boole at the Methodist college in Doncaster where he taught in the early 1830s, his petulant Unitarianism was already so well known as to constitute a possible block to advancement there, and twenty years later, his wife, Mary Everest, noted that he still adhered to this tendency.[50]

In compiling a work of interpretation rather than original research, where the intent was to bring the hero down from the realm of the abstract to that of the everyday, revealing the practical side of useful science that was the theme of the institute where he spoke, Boole portrayed a tinkerer Newton as well as a thinker: "The scientific adjustment of the paper-kite and its appendages, and the construction of sun-dials, water-clocks, and mill-work," he remarked, "afforded him a philosophical past-time, which could not fail to invigorate his natural powers of invention. Even his sleeping apartment is said to have been garnished round with the untutored productions of his knife and pencil." Once again perhaps with his own proclivities in mind, he added of the youthful Newton, "His mind appears at that time to have been much occupied with mechanical contrivances, yet not so deeply as to exclude the lighter amusements of drawing, and even poetry."

Just four years before Boole's speech on Newton, Sir David Brewster, Fellow of the Royal Society, published the first version of a substantial biography that eventually comprised two volumes, based on research into original sources including Newton's own papers that had gone out of the hands of the literary executor, John Conduitt, and into the collection of the earls of Portsmouth at Hurstbone Park (ending up eventually at Cambridge University).[51] Among the unpublished material was a wide range of autograph writings as well as notes compiled by Conduitt for a planned biography, eventually abandoned, but containing considerable amounts of precious firsthand information. A possible historicist revision of Newton's contribution seemed in the offing, but there were still numerous points that defied easy explanation.[52] Brewster was not immune to his own kind of mythmaking, and about the theological and alchemical manuscripts in the Portsmouth collection, which perhaps seemed to detract from the image of the scientist although they have been the object of a wave of recent scholarship, he said not a word in the 1831 volume, but then attempted to correct this omission in his definitive treatment in 1855 by explaining them away.

Concerning Newton's worldwide significance, Brewster's 1831 volume set off a noisy dispute among literati involving, among others, Jean Baptiste Biot, a pioneering astronomer. Biot wrote the entry on Newton in the French *Biographie Universelle Ancienne et Moderne*, a monumental publishing effort directed by Joseph Fr. Michaud and Louis Gabriel Michaud in Paris from 1813, and still consulted today.[53] Not enough adulation, complained Brewster. In the account of the priority quarrel with Leibniz about the discovery of infinitesimal calculus, and again in the account of the turn away from mathematics to other studies in the early 1690s, Biot failed to adopt the position of Newton's fiercest English partisans—an affront, said Brewster, that "has been the occasion of such deep distress to the friends of science and religion." Covering the whole dispute for the *Foreign Quarterly Review*, Thomas Galloway cried foul and summarized Biot's response to Brewster, published in the *Journal des Savants*.[54] Meanwhile, Whewell argued, against John Herschel (obviously with the Newtonian falling apple story in mind) that there are no really accidental discoveries. Only at a distance of a few years, perhaps in a fit of self-examination, Whewell began to wonder, At this rate, what would happen to the legacy of Cambridge's secular saint? If the wise did not provide the ignorant with the necessary heroes, who would?

Boole's approach did not fully gratify either side in this dispute. He sidestepped entirely the priority quarrel regarding calculus, only noting that in Newton's time, the method of "fluxions" existed "in a rude form, ... susceptible of indefinite improvement," adding, "but when the same instrument was applied to the same end by his disciples in France"—most recently, Joseph-Louis Lagrange (the Italian Enlightenment mathematician who pursued a career at the École Polytechnique)—"it had become doubly commodious and powerful, from the aggregate improvements of a century." Concerning the turn away from mathematics in the 1690s, he accepted the view attributing importance to an episode regarding the loss of manuscripts due to an accidental fire. Yet he shared Brewster's suspicion that this, and the ensuing mental impact on Newton, was no sufficient explanation. "It is unphilosophical to attribute to a temporary insanity, those religious impressions which so many other causes may have tended to develop."[55] Of the alchemical work, he followed Brewster's lead in saying nothing, also because he lacked the benefit of the Portsmouth manuscripts. And concerning the exegeses of biblical prophesies, studies of theology, and compiling of chronologies, in all of these he saw evidence of a mind preoccupied with the unsolved riddles of existence. Current work has vindicated this view.[56]

From the outset of his exposition, Boole distanced himself from the adulators. Even gravitation was no discovery of Newton, he claimed, referring instead to Plato, although the mode of proving it was. Hence he shifts the order of Brewster's topics, moving details (in chapter 17) from a letter to Robert Boyle discussing the effects of the ether, a material still considered at the time to constitute the space-filling element in the universe, back to gravity's first glimmerings (in chapter 11). Regarding the famous falling apple incident, supposedly told to Henry Pemberton at a distance of many decades, Boole expressed a sturdy skepticism. True or not, there was nothing here to lend credence to a mistaken belief that key discoveries may come easily. Rather, the homely trigger of a fallen fruit could have excited "some sleeping and folded energy" latent in the mind. Next came the struggle with erroneous calculations, trial and error, and only then, triumph.

The work on light, ingenious in itself, did not close the book on further explorations. "As a general theory of colour it is manifestly inadequate," proclaimed Boole. "Philosophers have been at some pains to point out its defects, though unfortunately they have been unable to find another half as general to supply its place." A great value of the work, apart from the conclusions, was the exposition of a method of generalizing from particulars. "It is true that his theory has been left imperfect; admit that in some of the applications it has failed. But at the same time we must acknowledge that in what he failed he did not fail as a common mortal, and that the marshalled intellect of Europe has vainly endeavoured to fill up the chasm!"

Although Boole clearly felt the weight of a past world of science from which he experienced some difficulty in extricating himself, nonetheless much had changed since the time of Newton. Scientific communication, for instance, had transformed from a regime of relative secrecy to one of relative openness.[57] In the seventeenth century, the notion of building a reputation through published work in specialized journals and timely volumes to be distributed to the learned world was still in its infancy.[58]

Francis Bacon's call for open research occurred when significant discoveries were still thought to be best protected by keeping them hidden and reporting them only to the tiny audience directly involved in the discoverer's advancement. Although Galileo pioneered in the use of printing to convey science to wider audiences, he also conveyed an early telescopic discovery to a correspondent by a clever enigma in order to protect it from rivals.[59] When Newton's first paper came out, *Philosophical Transactions* was only in its fifth year, and in 1675, he revealed the major experiments of the *Opticks* and his explanations of them in a lengthy manuscript sent to the Royal Society (itself a young organization), but only divulged them more or less entirely in print in 1704.[60] At first he kept the invention of calculus to himself and a few associates, such that considerable confusion later arose concerning who had it first. On this aspect of past science, Boole made no pronouncements.

Boole mainly sought to show research and discovery as dynamic processes in which Newton played a part. His own role, emulating the historian, would be to acquaint his audience with the latest insights, situating his subject in the context of another time.

ON THE CHARACTER AND ORIGIN OF THE ANCIENT MYTHOLOGIES

The quest for a key to all mythologies was not just the desperate project of a fictional character in George Eliot's novel *Middlemarch*. It was a genuine research program conceived by Victorian scholars attempting to balance the teachings of nascent ethnography and anthropology against the truths of faith. Boole, in this lecture, probably the same one listed as "On the Polytheism of the Eastern Nations" in the Mechanics' Institute records for the year 1841, refers to the work of Jacob Bryant, author of *A New System or an Analysis of Ancient Mythology*, which represents an interpretation of the accumulated research on the topic up to about 1776.[61] Bryant, who was also referenced in Eliot's 1871 novel set in 1829, thought to utilize an unconventional etymological approach to show that ancient mythological figures mostly stemmed from the Old Testament, with names modified in the process of transmission and translation. Boole was generally sympathetic, especially in regard to the primacy of the Judeo-Christian story. Still, he could not be so confident about the methodology. He may indeed have been alluding to Bryant's theory and the subsequent controversy when, in the lecture published here as "On the Character and Origin of the Ancient Mythologies," he said, "The attempt to reduce to some single principle the various and discordant relations of mythology is however little likely to meet with any success if we may judge from the past." He then added, "Such attempts are however not without their use," probably referring to the usefulness of error for arriving at the truth.

Bryant could have seemed like an example of probity in distinguishing fact from fancy, when not reaching for conclusions far too vast for the evidence at hand. Admitting that "we can not expect mathematical certainty; much less can we obtain experimental knowledge" in research on mythologies, since "the nature of the evidence will not admit of such a proof," he nonetheless insisted that "there are not wanting proper data to proceed upon." Thus, "we advance upon some sure grounds, proceeding from one truth to another, till we arrive at the knowledge required." Lines like these could have been written by John Locke. In fact, he went on to assert, the process of human understanding was essentially the same across the disciplines, including natural sciences. Just as "in the researches that we make in nature," he said, "some data are first stated; some determined and undeniable principles laid down, which are examined and compared: and then, by fair inferences and necessary deductions we arrive at the truth."[62] There was nothing in this approach that Boole could not endorse. Yet Bryant was only one of many then attempting to put research into humanity's religious past on a more scientific footing.

Some fifteen years before Boole would have spoken, Karl Otfried Müller, in a German work translated only in 1844, argued that unification theories were misleading and a methodologically sound historical approach called for finding the roots of each of the mythologies within the particular societies where they originated.[63] Now, unlike Casaubon in Eliot's novel, Boole could scarcely be faulted for refusing to take account of scholarship in the German language, to which he devoted even more tributes of admiration than he did to the French. He was also well aware that the long shadow of Enlightenment ethnography had begun to recede, and single-cause theories about the emergence of the first religions, ranging from Voltaire's notion of people's psychological need for an invincible leader to Nicholas Antoine Boulanger's proposal of a primal fear, were being replaced by conclusions drawn from deeper soundings in the local archives and a closer look at the archaeological evidence.[64] He refused, however, to relinquish the idea that Christianity and its biblical foundations were the unassailable bases of all beliefs, if only the necessary traces could be found to support the impeccable logic of the hypothesis.

In the very year Boole may be supposed to have been speaking, 1841, there appeared the *Dictionnaire égyptien en écriture hiéroglyphique* compiled by Jacques-Joseph Champollion based on the research of his brother Jean-François, whose publications on the Rosetta Stone throughout the 1820s allowed the first comprehensive insight into the meaning of Egyptian hieroglyphs.[65] Hitherto, guessing the significance of the various animal shapes and utensils, in their various contexts, displayed on various structures, borrowed, looted, or in place, had become almost a parlor game, and we find Boole engaging in it on occasion in this lecture. The comparison of the same text in parallel demotic, Ancient Greek, and hieroglyphic versions, copied from signs engraved on a monument captured first as booty by Napoléon's army in Egypt and again by the British army at the Capitulation of Alexandria in 1801, dispelled almost all previous conceptions about the supposed symbolic writing, including those most famously diffused by the baroque polymath Athanasius Kircher, but developed across the centuries into a largely fanciful repertory of conjectures tending to reinforce existing theories about the still-obscure form and structure of Egyptian culture.[66] With the discovery of hieroglyphs as a phonetic language, a new understanding of the ancient world and its expressions was almost within reach, although the path was not easy.

In Britain, the scholarly agenda regarding world religions was complicated by contact with a developing ideology about the imperial adventure.[67] Against Enlightenment figures like Constantin François de Chasseboeuf, comte de Volney, or Jean-Sylvain Bailly, who seemed to have been seduced away from Christianity by the magical allure of the pagan Orient, there arose a generation of writers for whom the history of religions provided a mental space and, eventually, platform for formulating the beliefs necessary to inspire a new generation of functionaries with confidence in the institutions as well as practices of the place from whence they came, along with a suitable share of diffidence toward the culture of the place where they were destined to rule. James Mill, father of the slightly less dogmatic John Stuart, drew on the hostile accounts of

William Ward and Claudius Buchanan, rather than the more conciliatory ones of William Robertson and Thomas Maurice, in the third (1826) edition of his *History of British India* to explain Hindu rites and rituals. There was much to be indignant about. "Though the progress of improvement has brought into comparative disuse the mode of seeking divine favour by the sacrifice of a fellow creature, horrid rites, which have too near an affinity with it, are still the objects of the highest veneration." Even the less violent observances seemed apt to create scandal: "A religion which subjects to the eyes of its votaries the grossest images of sensual pleasure, and renders even the emblems of generation objects of worship; which ascribes to the supreme God an immense train of obscene acts; which has these engraved on the sacred cars, portrayed in the temples, and presented to the people as objects of adoration ... cannot be regarded as favourable to chastity."[68] It was only to be wondered to what extent the British influence might bring the people to a more correct form of life. Boole shared this sense of outrage regarding the heathenism of both the past and present.

In fact, for his description of the Persian rites, Boole appears to rely mostly on Bryant, while keeping in mind Bryant's main source, Herodotus. He finds no fault, at least in this instance, with a method chiefly based on the literary records and hearsay, although even Bryant appeared to recognize the defects in, for instance, the account by the late seventeenth-century scholar Melchisédech Thévenot, including descriptions of tombs and altars supposedly encountered during a trip to Persia, but probably penned by a man who never left Europe.[69] Boole served his own purposes as long as the uncompromising otherness of what he called "that fearful worship" was sufficiently evoked, where "the frantic howls of exclamations with which they were attended, and the mournful character of hymns which were there chanted, gave a solemn terror to the localities and inspired" the devotees "with deeper and more superstitious reverence." For him, primitivism was the curse from which the peoples of the world were to be liberated by revelation of the truth.

For Boole, as for some of the scholarship he consulted, apparent uniformities were symptoms of a similar level of historical development across a wide range of cultures, and successive stages could be identified and traced. To be sure, the trajectory of material life in terms of the organization of production, from hunting to commerce, outlined by Adam

Smith, suited his purpose no better than the simple binary of savagery
and humanity posited by John Stuart Mill and others.[70] He may have
found something congenial in Hegel's exposition, published a few years
before he gave this lecture, of the four stages in the realization of freedom
through the emancipation of the spirit, going from oriental despotism to
the modern Christian state.[71] Rather than this, or indeed Giambattista
Vico's three stages comprising gods, heroes, and men, he adopts a tripartite structure somewhat reminiscent of Auguste Comte's, without the
culmination, envisaged by the latter, in a modern scientific point of
view.[72] In Boole's version, the primal adoration of nature would be followed first by the emergence of allegorical stories or myths, and then by
the development of a theodicy to account for the distribution of good
and evil, as follows: "In the first stage exhibiting the characteristics of a
very pure Druidism, the simple worship of External Nature; in its second
acquiring images and symbols, with probably some fragments of mythological history; in its third assuming the form of a philosophical allegory,
investing with personality those abstract ideas which had arisen in the
human mind, from an attempt to explain the seeming contradiction in
the moral government of the world."

Like the first sociologists of religion, Boole practiced a form of structural comparison by viewing cross-cultural uniformities as so many heu-
ristic tools for organizing a mass of evidence. Thus, "the fourfold division
of caste which is continued in India and Ceylon to the present day, which
anciently prevailed in Assyria, in Persia, and in Egypt, and of which in the
last country the traces are yet discoverable, was also noticed by the Spaniards as prevalent among the Peruvian and Mexican people." Again, "the
same may be observed of the doctrine of the Metempsychosis or transmigration of souls, of the privileges of the established Priesthood, of the
portion of soil allotted to their support, which in Egypt and in the American kingdoms was a third, and of various rights and customs as well
political as religious." Newton's *The Chronology of Ancient Kingdoms
Amended* is passed over in silence here as it was in the lecture on its author;
therefore there is no discussion about the conjecture that Solomon was
the first king in the world or Solomon's the first temple.[73] Nonetheless,
the basic inspiration behind the argument is rooted in theology. Boole
accepts Bryant's view that the biblical Deluge is the same single world-
changing event apparently recorded across a variety of early accounts. He

even extends to "the majority of classical antiquaries" an opinion that "the god Osiris" was "no other than the patriarch Noah, and his mythological history a corrupted form of the tradition of the Deluge." Likewise, "the Hindoo god Satyavatra [i.e., Satyavrata], whose history is undoubtedly taken from the Scripture account of the Deluge, or is itself a corrupted tradition of that event, is commonly represented as floating on an expanded ocean in the cup of the lotus."[74]

In regard to the relation between theology and morals, he clings steadfastly to the view investing monotheism with particular power as a civilizing force, in comparison to the alternatives. Here again, if not Comte, who decisively rejected this view, he could claim Bryant as an ally, but even more so, James Mill, for whom Indian religion exemplified a rude belief suited to a rude people.[75] Boole surmises that "under none of the forms which polytheism has ever assumed, or which it is conceivable that it might assume, is it capable of effecting any real or permanent improvement in the moral condition of the human race." Moreover, "while polytheism in all its forms is powerless to give effect to the decisions of natural conscience, its proper tendencies are invariably of a demoralizing character." Again, the imperfect development compared with European cultures is largely at fault: "While polytheism has often contained the scattered fragments of moral truth, derived, without question, from the unwritten dictates of natural conscience, it has never been able to give them their due force or effect, but in some instances has distorted them into an utterly opposite meaning."

This interpretation helps explain his indifference to, if not outright rejection of, the theses of Edward Gibbon concerning the rise of Christianity and fall of the ancient world.[76] If historical perspective for Gibbon did not imply relativism, so much the less so did it for Boole. Cynical administrators of the great despotisms may well have understood the potential of paganism for serving the needs of order and discipline, but they were looking in the wrong place, or at best, utilizing a faulty tool, compared with the more advanced religions of monotheism. Comparisons with modern peoples were instructive. "It is in the great communities of southern Asia, where man stagnates in hereditary bondage, that we see the worst results of Heathenism developed, in superstitions which are at once fanatical and gloomy, at once licentious and cruel." The British agents would clearly have their hands full.

Boole's beliefs about the civilizing mission of the empire were bound up with his beliefs regarding the saving mission of Christianity.[77] Religions were, in his mind, ridiculous not only to the extent that they diverged from a standard of reasonableness but also insofar as they diverged from the religion of Christ. In portraying Hinduism, for instance, Boole observed, "To Brahma the work of creation is assigned, on which he is supposed to have been engaged several millions of years. The method he employed is almost too absurd for description." The British administration is referenced as a conserver of religious order even among those sunk in superstition: "Between the two great sects, the worshippers of Vishnu and those of Siva, a deadly hostility subsists, even to the present day, and but for the dread of British power, armed hordes of fanatics might even now be carrying fire and desolation through the land." The abolition of the suttee and taming of the Thugs were part and parcel of the same move toward the necessary imposition of Western ways.

Laws might do much, but a people must be built from within. To explain the workings of moral codes in the best examples of human flourishing, Boole resorted to a metaphor from William Cowper's poem, "The Task." The passage in its full version (truncated by Boole) is as follows:

> Man in society is like a flow'r
> Born in its native bed. 'Tis there alone
> His faculties expanded in full bloom
> Shine out, there only reach their proper use.
> But man associated and leagued with man
> By regal warrant, or self-joined by bond
> For interest-sake, or swarming into clans
> Beneath one head for purposes of war,
> Like flow'rs selected from the rest, and bound
> And bundled close to fill some crowded vase,
> Fades rapidly, and by compression marred
> Contracts defilement not to be endured.[78]

Let there be a clearly defined allegiance and authority, in society and religion, Boole drew from this, and society will be an organic whole rather than an ad hoc collection of discontinuous bodies, which historian and moralist alike, observing primitivism in all its forms, including those manifestations evident from time to time in the West, must regard as a monstrosity.

ARE THE PLANETS INHABITED?

On an evening in February 1844, Boole gave his lecture at the Lincoln Topographical Society on a subject that he had already briefly mentioned in the lecture about Newton: namely, the possible existence of extraterrestrial life.[79] On that previous occasion of some nine years before, he had lauded the great mathematician for contributions to the science of astronomy. "Those bright and distant worlds, whose laws it was reserved for him to investigate," Boole had instructed his audience, "have ever been the objects of human curiosity." In due course, superstitious beliefs about the power of stellar forces to rule human destiny had been replaced by "a better philosophy." As the great discoverer of universal gravitation knew so well, however, modern astronomy by no means diminished the awe and wonder the planets might inspire. Now they were viewed "as individually possessing an interest of their own, the abodes of other forms of organic life, of other orders of intelligent existence." In this guise they now excited the curiosity of many, including Boole himself. The discussion about the plurality of worlds, or pluralism, was fast becoming one of the passions of the time.

"Such are the prospects of modern astronomers," Boole had concluded his assessment of pluralism back in 1835: "bold, yet scarcely conjectural." In his enthusiasm, perhaps, he somewhat overstated his case. And still in 1844, the topic was highly controversial, in spite of what seemed like a media obsession. Scientific evidence supporting claims about pluralism was by no means conclusive, and theological problems abounded in relation to Christianity. The evident inconsistencies seemed to put undue strain on even the loosest allegorical readings, if the Garden of Eden story in Genesis was to indicate a real starting point for life. Moreover, how to understand salvation if humans were not the only intelligent creatures? Such questions had addled the minds of the best thinkers of a previous century, and if the Enlightenment philosophers conceived of varieties of deism and even materialism capable of accommodating speculations on pluralism, this certainly did not necessarily endear them to orthodox believers or scriptural literalists.[80]

Nevertheless, some forms of once-radical thought were becoming widely accepted by Boole's day, and pluralism to some seemed no more outrageous than the notion that social and economic progress must inevitably be accompanied by intellectual regeneration. Protagonists of the

Romantic movement sensed a changing intellectual vocabulary even when they did not embrace the dogma of progress, and regarding pluralism, Percy Bysshe Shelley, Lord Byron, and William Wordsworth all weighed in, dedicating serious and pithy lines to the matter. The latter referred already in his 1798 poem "Peter Bell, a Tale" to "the red-haired race of Mars" and "towns in Saturn."[81] If contemplation of outer space served in these cases as a counterpoint or inspiration for the continuing quest for worlds within the mind and soul, the same perhaps may be said of Boole.

So compelling was the universal presumption of truth in regard to the existence of extraterrestrial life that just a few years before Boole spoke, a New York–based journalist named Richard Adams Locke poked fun at the earnest claims being advanced between Europe and the United States, even by scientists and theologians who ought to have had specialized knowledge to make definitive pronouncements. In an 1835 edition of the *Sun*, he published supposed drawings of observations made, so he said, by an assistant to John Herschel, the science methodologist, using a telescope located at the Cape of Good Hope and directed at the moon, showing winged humanlike creatures in flight, presumably inhabited buildings, and beasts of various kinds. Many who read the article were reportedly disappointed when "the Great Moon Hoax" was eventually unmasked.[82]

Perhaps because of the debate's paradigm-breaking fusion of wild speculation and pseudoscience, combined with persistent prejudices about who was qualified to speak the truth, numerous theologians contributed. To be sure, whether the Scots clergyman Thomas Dick ought more accurately to be classified as a theologian, philosopher, or astronomer is perhaps beside the point. In his *Celestial Scenery* (1836), one of a series of books on religious and astronomical subjects, he listed what he thought to be the populations of Mercury (8,960,000,000), Venus (53,500,000,000), Mars (15,500,000,000) and so forth, even including lately discovered Uranus (1,077,568,800,000). For Dick, it was an effect of God's providence that the universe should be filled with God's creatures, including those made after his own image along with those decidedly not.[83] Boole would agree.

John Pringle Nichol, whom Boole cites in the lecture, was among many serious contemporary scientists who embraced pluralism with no compunction. There was plenty of good precedent. Technological advances in stellar astronomy had sparked new interest in a question posed at least as far back as Aristotle. Would there finally be a definitive answer? William Herschel, John Herschel's father and the discoverer of Uranus, held out such hope until his death. Nichol accordingly accompanied his account of the solar system with suggestive drawings presumably representing telescopic observations, including some influenced by the work of the German astronomer Johann Hieronymus Schröter showing the planets' phases and uneven terrain. "But is Life in all these planets?" he asked. "Through all possible schemes, through all conditions of a Globe's evolving organization, is what we call Life an inseparable or essential concomitant?" Taking the evident presence, here and there, of an atmosphere, he made a suggestive case. "Life, visible or invisible—i.e., the sentient and intelligent principle—nay even, progressive life, a growing and evolving Reason—is doubtless an essential element of the Universe."[84] From such a hypothesis to full-bodied pluralism was only a small step, and Nichol was not afraid to take it.

Boole, too, followed what he believed to be the logic of current observations and interpretations. Taking for granted the operation of an intelligent maker, he argued from the design to the function of what appeared. To explain his approach, he reverted to the example of biology. "From the fossil remains of extinct animals," he recalled, "Cuvier inferred their forms and habits." The method was firmly based in scientific principles. "The confidence which we attach to the results of such inquiries rests on this," Boole explained: "that we believe there are evidences of design even in the structure of a bone, or the convolutions of a shell, and that the object of the design may be inferred from the uniformity of general principles by which all the Creator's works are characterized." Were not the planets all predisposed, in the construction of the universe, so as to receive solar warmth alternately on various parts of their surfaces, just as the earth? Did they not all appear to possess an atmosphere and such terrain as would be most apt for collecting water? Then the three chief conditions of life on Earth might plausibly be met. He stopped short of advancing more claims than he thought his evidence might bear. But like biology, astronomy was a science of hypotheses as well as deductions, and a good theory might begin from a conjecture arrived at by proceeding from what was known to what was not known. Keeping in mind the guiding theme of intelligent design, the conclusion was inescapable. "If it is granted that the Author of nature acts consistently in all his works, and that he accomplishes the purposes of his will by special means and adaptations, that his purpose with respect to the Earth is, that it should be an abode of life, and that the mass and adaptations by which that purpose is accomplished are employed in other worlds, I see not how the conclusion is to be resisted, that those other worlds are intended to be habitations of life also." The opposing view at this point seemed nothing less than blasphemy. "Refusing to admit the inference, we fall on such contradictions as these: that he acts inconsistently whose character is immutable; that he has created worlds without end or object, who made nothing in vain; that he has chosen to preside over an empire of death, who is the source and fountain of life and enjoyment." And a world without God was as preposterous as a science without God.

To steer the course of faith along lines that made sense from the standpoint of observation, experience and logical reasoning was the leitmotif of his career.

A PLEA FOR FREEDOM

Based on a few hints below, we deduce that Boole may have made these brief remarks in 1846, around the time when he was asked to give his views on library collections, a year or so after James Snow, MD, sometime mayor of Lincoln, assumed the presidency of the Mechanics' Institute following the resignation of founding president Edward Bromhead. Controversy was in the air. The institutes had been set up on the principle that discussions about religion and politics would be excluded from normal institute activities, also because elsewhere the issues connected with these topics had been tearing British society apart. Should the prohibitions extend to reading?

To be sure, the 1840s were no longer the 1820s, and the laws against printing and circulating material on current politics, enacted to prevent a repetition of the 1819 unrest that ended in the Peterloo Massacre, were not so often enforced. Already in 1825 they were considered as obstacles to the free circulation of ideas on which public education depended. Henry, Lord Brougham, an early patron of the London Mechanics' Institute, in a speech on the topic, pointed out that the causes of religion and politics alike were better served by freedom than by censorship. "The great interests of civil and religious liberty are mightily promoted by such wholesome instruction; but the good order of society gains to the full as much by it."[85] Boole agreed. And of the role of public libraries in the diffusion of knowledge he could speak from personal experience. His own intellectual itinerary depended on public access to books, often provided by the Lincoln Mechanics' Institute, whose mission included public education through reading. Indeed, the precincts of the institute library had actually been his home for the period when his father, John, served as curator there with a small salary. In time, he would be asked to survey the quality of texts as the library developed its collections on different subjects.

In his address, Boole referenced the dual prohibition, within the institute, of political and religious discourse, but his real concern was with the latter. He would have agreed with John Milton, a writer whose works he knew well, and who expressed in the *Areopagitica* that the truths of faith were only strengthened, never weakened, by the exchange of ideas.[86] And contrasting ideas abounded in this period. The 1820s had seen the repeal of penalties for dissenters from the Church of England and even the granting of political rights to Catholics. Now the religious landscape in England included a vast array of tolerated sects, and the church was in constant disagreement with itself.[87] In the absence of any certainty regarding the best route to salvation, discussion about religious issues, Boole believed, was the only route to truth.

His own religious convictions are somewhat hard to specify. The vague affinity for Unitarianism evinced from time to time seems not to have developed into any definite affiliation. In his remarks here he flatly says, "I am not an Unitarian." Similarly he asserts, "I have no moral sympathies with what is called Calvinism." In spite of his nominal attachment to the Church of England, he distanced himself from the Oxford movement, a group of fellows at Oxford's Oriel College considered by some as bent on returning to Roman Catholic roots. He was well aware of the movement's efforts since the early 1830s to direct critical attention to the very bases of the English Reformation, inciting debate about the significance of the Elizabethan Religious Settlement, the concept of the consecration of the Eucharist, and much else. Just a year before the presumed date of the remarks published here, John Henry Newman, leader of the movement, actually converted to Romanism, leaving Edward Bouverie Pusey to take his place. "I dislike Puseyism," Boole exclaimed, adding in an ecumenical spirit, "but I would permit it to speak for itself."

The other figures mentioned by name below testify to Boole's wide reading more than to his religious preferences. Archibald Alison, a Scottish Anglican priest, wrote books of sermons on subjects such as "On Spring" and "On the Encouragement to Active Duty," and is also known as the author of a theory that tried to explain aesthetic experience in terms of an individual mind's mental associations.[88] Edward Thomas Vaughan, rector of a parish in Leicester, wrote about Sabbath observance and Christian benevolence, in the latter work expressing support for the policy of teaching trades to "poor girls, who, being trained up in ignorance and idleness, and exposed in early life to the contagion of bad examples, are unfitted for any useful office in society, and often fall a prey to seduction and prostitution."[89] Thomas Arnold, historian, educator, and theologian, devotee of Broad Church Anglicanism, wrote *Principles of Church Reform* in 1833, maintaining an Erastian position against the High Church, and in 1841, the year before his death, he became Regius Professor of History at Oxford on the merits of a multivolume *History of Rome* begun in 1832 but never finished.

What Boole may have found in the "unrivalled dissertations" of Jonathan Edwards is worthy of speculation. The North American theologian and eventual president of the College of New Jersey (Princeton) struck a distinct figure from those already mentioned. A powerful proponent of the Great Awakening, he used his sermons to stoke the movement of religious revival that swept North America and Europe in the 1730s and 1740s.[90] Drawing on New England's Puritan heritage, he encouraged a return to individual conscience and a highly personal form of faith, against constituted religious authorities. His most notable sermons attempted to engage the emotions on issues of faith and salvation, such as in the typical text "Sinners in the Hands of an Angry God" (published in 1741), proclaiming the real immediacy of God's intervention in human lives, the benefits of faith, and the wages of sin. Other works delved into church history and traced the course of missionary activity among North America's native peoples. All were included in the latest two-volume 1840 edition of the *Works*, prefaced by "An Essay on His Genius and Writings, by Henry Rogers."

Everest later described Boole's outlook as involving a fundamental religious feeling without any subscription to a particular sect. "Though he did not exactly believe doctrines," his wife said, "he seemed to care for them in a way that I was never able to understand." The distinction was highly significant. She went on, remarking, "He used to say that if the doctrine of the Trinity were in any sense true, it must be in this sense: not that there are three persons in the godhead, but that man is so constituted as to be capable of comprehending three, and only three, modes of manifesting of God." She added, voicing her customary skepticism about all faiths, "I often asked him why people seemed to suppose that God would be lost to them unless he had at some time walked about in human shape and performed miracles." Whether his answer "that that was just what he had never been able to discover" was spoken simply in a spirit of lighthearted repartee or not, we cannot determine with certainty from the writings. Most likely, Boole's religious ideas evolved significantly over time and underwent numerous shifts. Already in 1840 he told a correspondent that he hesitated "to avow myself in belief a Christian," while nonetheless placing his "hopes of future happiness on the great propitiatory sacrifice and … merits of the saviour." He emphasized, "I doubt whether I am a Christian at all except in mere speculation." Later, possibly also influenced by the decidedly secular Everest, he would affirm, in her words, that "he had no positive belief in a future life."[91]

Still, once again in Everest's words, "The hope of his heart" was "to work in the cause of true religion." Indeed, somewhat more surprisingly, "mathematics had never had more than a secondary interest for him; and even logic he cared for chiefly as a means of clearing the ground of doctrines imagined to be proved, by showing that the evidence on which they were supposed to rest had no tendency to prove them."[92] Such a statement seems to ring true, especially in light of Daniel Cohen's recent interpretation, demonstrating just how closely Boole's scientific ideas were bound to his religious ones. For instance, according to Cohen's account, in formulating a new science of mathematical logic, Boole derived his main insights from a particular view regarding the mind's innate sense of unity planted there by a Creator. Moreover, many of the laws of thought (as he called them) were cast in terms intending to add precision and clarity to the basic issues in religious controversy, exemplified in his (supposed) analyses of the theological principles of Benedict de Spinoza and Samuel Clarke.

Whatever may have been his own convictions, he considered that readers at the Mechanics' Institute and anywhere else ought to be allowed to read about any creed. If his examples in this discussion include only Christian authors, this does not necessarily suggest that he would have objected to including religious works proposed, for instance, by the Jewish scholar who he once claimed to have set him on the way to philosophical studies at an early stage or even such Islamic writers as he may have had at his disposal. The fundamental flaw in the restriction, he believed, was not so much in the banning of specific doctrines as in a misunderstanding of the basic unity of all knowledge, of which theology formed a part, according to his epistemology. Since the greatest philosophers all touched on theological ideas, banning from the library whatever contained such notions was equivalent to banning much of philosophy. Boole himself was no different in this respect from Locke, Smith, and Isaac Barrow, all of whom referenced theological concepts in their work.[93] In this lecture, he might have added the name of Newton, whose theory of planetary motions, like his own view of the laws of thought, was supposedly founded on a fundamental concept of God's providence.

A word or two is in order regarding the style of the text. Evidently we have here a set of notes (written out by Boole himself in fine copy) for an oration to be given before a select audience of committee members at a policy meeting to decide on the future of the Lincoln Institute. Therefore, in comparison to other texts in this book, the sentences are shorter and the reasoning more concise. We may imagine that the filling out of certain expressions of particular terseness would have been saved for the moment of extemporaneous delivery. Still, we see Boole's mind at work in some typical patterns, and there are specific insights here that do not appear elsewhere.

THE RIGHT USE OF LEISURE

The year 1847, when Boole gave this speech, was marked by triumph for British labor and social movements, but also by frustration. Parliament passed the Ten Hours Act restricting the working hours of young males ages thirteen to eighteen as well as women of all ages in textile mills to

ten hours per day. The struggle for factory legislation had been long and
bitter, and was not over yet.[94] Similar bills had been defeated twice in the
previous three years on various grounds including fears regarding a reduction of profits and the threat of foreign competition. Even a twelve-
hour bill was passed in 1844 only amid an uproar. The proposition to
extend time restrictions to adult males, still not decided, was fraught with
another set of difficulties connected with the liberal dogma that individuals should be free to choose long hours if they so wished.

Labor leaders and educators alike added their voices on the issue of the
working day. Whig politician Thomas Babington Macaulay couched
things in patriotic terms: "Rely on it that intense labor, beginning too
early in life, continued too long every day, stunting the growth of the
body, stunting the growth of the mind, leaving no time for healthful exercise, leaving no time for intellectual culture, must impair all those high
qualities which have made our country great." On the other hand, "a day
of rest recurring in every week, two or three hours of leisure, exercise,
innocent amusement or useful study, recurring every day, must improve
the whole man, physically, morally, intellectually; and the improvement of
the man will improve all that the man produces." This he contrasted with
reports regarding the intense exertions of laborers at the Krupp works in
Germany, phrased in characteristically jingoistic language: "My honorable
friend seems to me, in all his reasonings about the commercial prosperity
of nations," he said, in his 1846 address to Parliament, "to overlook
entirely the chief cause on which that prosperity depends. What is it, Sir,
that makes the great difference between country and country? Not the
exuberance of soil; not the mildness of climate; not mines, nor havens, nor
rivers. These things are indeed valuable when put to their proper use by
human intelligence: but human intelligence can do much without them;
and they without human intelligence can do nothing."[95]

Mainly concerned with urban store employees, perhaps benefiting
from the natural prejudice among social improvers in favor of a class that
was not so abject as the factory-employed counterpart, was the movement for "early closing," which of course also implied shorter hours for
workers. Generated by an Irish journalist and activist named Samuel
Carter Hall along with others in the 1830s, the movement began to catch
on in the following decade.[96] To his consort, particularly concerned with
the repercussions for women, Hall attributed the expostulation, "The

Late Hour Employment in Shops is proved, beyond controversy, to be needless for any beneficial purpose, either to buyer or seller. It is oppressive and cruel as well as unnecessary." She went on (Hall reported) to assert, "It condemns many thousands of industrious persons to that 'excessive toil' which destroys health, and retards or prevents religious, moral, and social improvement. Out of it arise innumerable evils, and no single good: debilitated constitutions, impaired minds, absence of religious thoughts, ignorance of moral duties, or inability to perform them—are but some of those evils. Over-work is the sure passage to an early grave—for which there has been no preparation."[97] While by 1846 the London branch received the formal denomination of the Early Closing Association, also by that year a Lincoln branch had been founded, involving Boole.

Even as debate raged concerning the new proposals on factory hours, circumstances intervened to force a decision. A worldwide economic downturn from 1846 seemed destined to extend the threat of mass misery, already endemic in Ireland, to England as well.[98] A major principle of the outgoing conservative ascendancy was shattered by the repeal of the Corn Laws in 1846, allowing the temporarily cheaper continental grain to feed the hungry people of Britain, over the howls of the less clairvoyant local producers, but somewhat to the satisfaction of the leaders among the manufacturing middle class who thought they might savor yet another slight against the great landed interests who in their view stood in the way of economic progress.

With the Chartist movement gaining ground, and calls for genuine electoral reform becoming too loud to drown out by half measures such as the reform bill of 1832 bringing the total number of voters up to some eight hundred thousand within a population of fourteen million, the winds of revolution once again sweeping through Paris and the rest of Europe blew harshly in Britain too.[99] The most strategically minded among the liberal Whigs contrived to persuade other political forces that the only way to avoid admitting some form of real democracy was to concede on the less critical problems—namely, conditions, lifestyle, and possibilities for self-advancement, which were just the sorts of issues on which the Young Men's Christian Association, Early Closing Association, and ten-hour agitators all agreed.[100]

Whatever freedoms or constraints on working hours applied to manu-facturing according to the Factory Acts, as far as local retail store hours were concerned, cities were allowed to make their own rules. Boole hints at the beginning of his talk that such rules had recently been imposed in Lincoln, giving some cause for celebration. As had already happened else-where and would continue to happen around the country, the municipal government had set the closing hour back a notch or two for consider-able sectors of the urban retail workforce. Shorter work hours for numer-ous apprentices, shop assistants, and assorted retail personnel seemed to imply longer hours at other activities, and here the new legislation, with hopes for more to come, suggested opportunities for directing popular energies to useful purposes. What might these be? The topic excited the imaginations of scores of well-meaning orators, and Boole, as always well prepared not only by his studies but also by his life, rose brilliantly to the occasion. In his view, as in that of a wide range of early closing advocates, the usual activities through which a small ugly core of workers had falsely colored impressions about the rest were studiously to be avoided—that is, loitering, drunkenness, brawling, gambling, and general dissoluteness. What to put in their stead?

While divines hoped for a new influx of the faithful into church-related activities, among popular leaders, the cause of leisure time met the cause of popular education.[101] A prize essay for the Metropolitan Drapers' Association in 1843 declaimed against *The Evils Which Are Produced by Late Hours of Business*, offering excerpts from the debates regarding the 1832 Factories Bill, demonstrating that much of the work of social improvement was left undone. Here Boole's engagement in the Mechan-ics' Institute coincided with the matter at hand.[102] "There are so many things which it is good to learn, so many which it is good to do, as to afford employment for all our faculties both of thought and action. How-ever various our tastes, our dispositions, or our powers, we have before us an ample field for their exercise." Let there be books, study, and conversa-tion regarding the wider issues of life and thought; let there be sports, physical activity, and whatever might push the whole human being in a virtuous direction. To be sure, pursuits tending in any other direction were utterly "without excuse."

He easily dismissed the notion that intellectual curiosity might in some way be sinful, which apparently was still uttered in great seriousness within some of the more austere quarters of the religious establishment, judging at least from the anxious ridicule poured on such a view by the Manchester Baptist minister William Gadsby in a letter to the *Sunday School Teachers' Magazine* in 1846.[103] Ages-old prejudices still commanded significant support among the sizable conservative elements of society, and this is what Benjamin Disraeli assumed when he satirized them in the 1845 novel *Sybil* by making the character Lord Marney express the hope that there will be no infants' schools in his neighborhood due to the possible mischief that might result from excess instruction.

Not surprisingly, debates on education in the 1830s and 1840s focused as much on what was to be taught as on whether there was to be teaching. Should efforts be directed by the state or left up to the communities? Should the Established Church be in charge? Important religious interests were involved. A sector of the newly emancipated nonconformist groups earned a reputation for obscurantism by opposition to the education clauses of the 1843 Factory Bill, such that it turned to the public press to correct the misapprehension that the target of its scorn was education in general, when in fact it wished to reprehend only education that excluded its ideas. "The injustice of the reproach ... is proved by the extremely unjust and dangerous provisions of the Bill itself [and] ... second, by the unparalleled unanimity with which all religious denominations combined to repel the attempted usurpations of the Established Church."[104]

Far more than the bickering between sects, what interested Boole in the matter of leisure time and self-improvement was the issue of knowledge per se along with the modes of acquisition and verification. Some of his thoughts may well have gone over the heads of most listeners, although various combinations of them would turn up in his published works. A major concern in the lecture below was the question about the subject and object of the mind's operation.[105] "Truth," he says, is not to be regarded as "the mere creature of the human intellect." Instead, it led an independent existence. Likewise nature too was external to humans and knowable. Hence the idealism of George Berkeley and the antinaturalism of Immanuel Kant were equally misguided. Nor was the theological concept that divided human from divine knowledge to be believed unquestioningly. Both types were one and the same.

His exposition of the purely intellectual side of virtuous leisure entailed an account of past and current contributions to various fields, revealing much about his own personal preferences. In historiography, the quality of rhetoric and exposition were decidedly of secondary interest, so long as the doctrine reflected recent convictions and spoke to a modern way of life. Thus there was no Herodotus, Julius Caesar, Livy, or Tacitus; there was no Francesco Guicciardini, Niccolò Machiavelli, nor even Milton. Two exceptions to the rule of preferring modern scholarship to historical sources were the *History of the Rebellion*, an account of the English Civil War from 1640 to 1647 by Edward Hyde, the earl of Clarendon, a participant in the events, and Voltaire's *Age of Louis XIV*.[106] Possible explanations may lie in the particular doctrines enshrined in these two works. The first, though written from a royalist (and not a parliamentarian) point of view, was far too honest about Charles I's foibles to be considered unnecessarily laudatory or flattering, while at the same time showing a contempt for incendiaries and wanton rabble-rousers that even the staunchest Whig could only have found congenial in the explosive climate of the 1840s. Voltaire's work was a paradoxical appeal to the best examples of authoritarianism at the service of peace, prosperity, and the arts by a leader of the French Enlightenment.

The account of history as a field of discovery and interpretation for furthering the moral development of people reflected a lifelong passion. Inquiry here, in Boole's view, proceeded progressively to greater knowledge, so he chose recent accounts, even when written by those better known in other genres, such as Friedrich Schiller, included here not for poetry but for the *History of the Thirty Years War*, which begins thus: "Since the commencement of the religious war in Germany, until the peace of Munster, scarcely any thing great and remarkable took place in the political world of Europe in which the Reformation had not the principal share." There could be no doubt about the fact or its significance: "All historical events which happen in this time connect themselves with the reform of religion, if they do not originally flow from it; and all states, of whatever importance, have experienced, more or less, mediately or immediately, its influence."[107] Other European history we find represented by Simon de Sismondi and François Guizot. Britain is covered by David Hume, a founder of the "Whig Interpretation," and also by George Lillie Craik. In this last connection, the explicitly referenced *Pictorial*

History of England, with Craik as coauthor, points to the London-based "Society for the Diffusion of Useful Knowledge," dedicated to supplying new readerships with inexpensive texts—a program fully aligned with Boole's interests.

Choices include such scholarship as sought to submit historical sources to the new diplomatic methods just then being introduced in the German universities. Barthold Georg Niebuhr, referenced in this lecture, acknowledged the benefits in no uncertain terms: "The end of the last century was the opening of a new era for Germany," Niebuhr informed. "Men were no longer satisfied with superficial views in any field of knowledge: vague empty words had lost their currency; but neither was the work of destruction, in which the preceding age, indignant against protracted usurpation, had taken pleasure, any no longer held to be sufficient." A pattern had been set for regular professional exercise in the fields of scholarship. "My countrymen strove after definite and positive knowledge, like that of their forefathers," Niebuhr explained, "but it was after true knowledge, in place of the imaginary knowledge which had been overthrown." The example of his own discipline was particularly illuminating, especially in regard to the emphasis on history as research rather than as a an unchanging block of notions: "If a previous age had been content with looking at ancient history in the way many look at maps or landscapes, as self-sufficient, rather than as the indispensable means for bringing a picture of the objects in question to the soul's attention, now [ancient history] could no longer be considered satisfactory unless the clearness and distinctness were comparable to the history of the present age." The particular political and social context of Europe had conspired to produce these results. He went on to explain that "the time was one in which we witnessed many unheard of and incredible things; when our attention was attracted to many forgotten and decayed institutions by the sound of their downfall; and our hearts were strengthened by danger, as we became familiar with its threats, and by the passionate intensity given to our attachment to our princes and our country."[108] In the same spirit perhaps, Boole cites work by Arnold Hermann Ludwig Heeren, professor at the University of Göttingen and, for a time, editor of the *Göttingischen gelehrten Anzeigen*, a major scholarly journal, and contributor there of various articles presenting inquiries regarding the sources of the ancient historians and geographers.[109]

Other writers in the history section of the speech are as interesting in themselves as they are for what they may reveal about the speaker. Alexander Fraser Tytler is known for a particularly harsh view of pure democracy, not in the work mentioned below, but in the Athens section of his *Universal History*, a compilation of his lectures on history at the University of Edinburgh, published posthumously in 1834.[110] William Cooke Taylor, born in Youghal, County Cork, apart from his anti–Corn Law activism, observed factory operations in England for an 1844 book called *Factories and the Factory System* dedicated to Sir Robert Peel. There he proposed to weigh the claims for and against the system by the criteria of economics, which in his view bore "the same relation to social questions that Mathematics do to Astronomical phenomena." The conclusions were basically favorable. The "manufacturing industry," he said, "associated with commerce, is the only source of the wealth and greatness of nations," whereas places like Ireland that mainly produced raw materials were destined to poverty. The multiplication of useful articles for consumption and competitive struggle between owners to gain the advantage of the latest mechanical inventions produced by the brightest minds far outweighed the temporary setbacks caused by the business cycle, or attempts on the part of workers or owners to manipulate the market. For an example of the factory system in operation within a context of social peace, he pointed to New England, where unlike in Britain, a vast reserve army of unemployed hands did not depress wages.[111] On the whole, in this view, industrialization left as profound a mark on modern economies as the Protestant Reformation had on religion, and the world would never be the same again.

Absences may be as significant as presences. There is no trace of Gibbon's *History of the Decline and Fall of the Roman Empire*. The elegantly written work, published between 1776 and 1788, signaled a new approach combining philosophy and philology—perhaps not new enough, by Boole's time, and perhaps the emergence of the German school impressed him more. On the other hand, the work's Enlightenment imprint may have struck him as old-fashioned or even transgressive. Controversy still smoldered regarding the notorious fifteenth and sixteenth chapters, blaming the advance of Christianity for the decay and disappearance of much that was great in ancient civilization, and focusing, with a lavishness easily mistaken for voyeuristic glee (so said the critics), on the persecutions of the Christians in the first centuries. Just in 1838, Henry Hart Milman, an Episcopal priest, had produced a new annotated edition acknowledging the value of the work, but attempting by copious reproaches to attenuate the possibly damaging effects of the offending chapters in the interests of true religion and the "moral dignity" of the faith to which Gibbon's "imagination is dead."[112]

The study of nature, much more than history, directed inquiry, in Boole's account, to laws and regularities governing the multitude of particulars. He is strangely silent regarding current practitioners in the fields of natural knowledge, preferring to elucidate the method rather than describe its users. Accordingly, he outlines the procedures of inductive reasoning for arriving at general principles and deductive reasoning for predicting new phenomena. Astronomy, among the other sciences, in his view has achieved the highest stage of perfection, so that new things may be abundantly discovered by drawing out the consequences of, say, the law of gravity, in light of the observed data. In one case alone, he refers specifically to recent work: the case of the discovery of a new planet, given the name of Neptune in this same year, 1847.[113] He omits the priority dispute between Urbain Le Verrier and John Couch Adams regarding the discovery, just then decided by the Royal Society in favor of Le Verrier. Yet the case is ideal for his point related to predictability, where observers noted aberrations in the orbit of Uranus unexplainable by the sole effect of the sun and its own motion, supposing consequently the presence of another object (i.e., the new planet) exerting gravitational force, which the optical evidence bore out.

In spite of the power of the method and range of recent discoveries, science was far from realizing its promise as an agent of positive change. It might do so, Boole thought, if only people could be persuaded to pay more attention to it, but in this the Mechanics' Institutes had failed so far. Other studies surpassed it in influence and prestige. Part of the problem might lie in the inherent difficulty of the associated concepts, he admitted, or even a general public indifference to the kinds of things that comprised it. "The degree in which it ought to be made an object of pursuit, must depend upon taste, leisure, and ability." This was no reason for despair, however. On the whole, he suggested, there were other areas of study that could contribute at least equally to human improvement— studies concerned more directly with the way people lived. "Man must ever be the great study of man."

Joining the study of human behavior with that of mind was the field of moral philosophy, and here is another area to which Boole devoted particular attention. The acceptable authors were many, and the problems were tough. For a layperson's introduction, he found Francis Wayland's 1835 book *Elements of Moral Science* "well-deserving of attentive perusal." Written by a former physician who had gone on to become president of Brown University, it claimed to improve on the long-used manual by William Paley. In place of Paley's fundamentally utilitarian approach, it offered a view that allowed ample space for the operation of religious convictions. The Benthamite principle of the greatest happiness for the greatest number, it insisted, was no sufficient guide to moral action, either for an individual or the group, in a supposedly Christian society. Instead, the agreement or disagreement with divine revelation must be taken into account, and righteousness was always to be chosen over pleasure.[114] Boole commented, "Rectitude of intention, and an earnest desire to carry into practice the truth to which we have already attained, are in some measure necessary to correctness of judgment."

In the search for models of probity and style, the authors to be avoided were as significant as those to be followed. Ralph Waldo Emerson and Thomas Carlyle, for instance, were condemned, mainly due to their tendentious mode of exposition. Bad enough was that their "style and manner betray an imitation of the German model," which apparently suited writing in German but not English. Worse yet, "they are deformed ... by frequent extravagance of language, startling paradoxes, and positive error and self contradiction." Thus, even though "we find in their writings many noble thoughts, much force of language, and an apparent earnestness of spirit," there was too much worthy of reproof. "On the whole, I conceive, that this school is built upon a false foundation." Boole despised the "half idolatrous veneration with which they are regarded by their disciples," considering that "the investigation of truth is too solemn, too difficult a thing, to allow of its being associated with a constant effort to appear striking and original."

The best teacher of all (and this is perhaps the chief message of the speech) was nature itself. Not only from the standpoint of the information that might be abstracted in order to form conclusions increasing the sum of knowledge in natural science. That came from the operation of the scientific method. Also, nature, if only contemplated where existing in abundance, supplied moral instruction of a kind unavailable in an urban setting.[115] Like Smith and other moralists who viewed agriculture as the prime source of value—the real "wealth of nations"—and agriculturalists as the most solid basis for a just and plentiful society, Boole was profoundly diffident about the effects of city life on the human spirit and culture in general.[116] "We are apt to become too artificial, to conceive of too many things as necessary to our happiness, to forget the simplicity of our real wants in the appliances of luxury, and the grandeur of our real nature amid the external trappings of society." Experience of abundant nature was therefore essential since "intercourse with the works of Nature tends to correct those false and exaggerated views of life, which a long residence in cities is calculated to engender."

It was a powerful theme, and one that had become particularly prominent in European thinking in Boole's time. The pedagogical naturalism of Jean-Jacques Rousseau, whose fictitious protagonists went out into nature for moral inspiration, deeply impressed itself on Johann Wolfgang von Goethe and later Johann Heinrich Pestalozzi in spite of Kant's insistence on the categorical imperative as the foundation of morals.[117] On Boole too. What were the lessons to be learned? "There is a silent force in the appeals of Nature to the soul of man," said Boole, "and that force is always exerted on the side of Virtue." The teaching could be as simple as an understanding of human behavior in relation to other living things, such as in the allegorical references to social insects in the moralizing work of Bernard de Mandeville, or a reminder about the cycles of birth and death. Boole quoted a passage from Alexander Pope's translation of Homer's Iliad: "Now green in youth, now withering on the ground." Or the teaching might inform about duty, beneficence, beauty, and the greatness of God. No wonder the experience of nature was such a "powerful restorative to the mind that has been overtaxed with labour, or worn with care."

In spite of the frustrations involved in making a message of moral regeneration heard in an epoch of economic catastrophe and political crisis, Boole remained optimistic (or so he said) regarding the good intentions of his listeners. Thus, he urged them, while contemplating leisure time activities, to consider what portion might be devoted to improving the condition of their fellow humans less fortunate than they. A little effort, a small sacrifice, might mean much to one bowed down by indigence and misfortune, and the rewards would be great. Let the members of his audience bring themselves to dedicate some portion of their time and effort to the benefit of their fellow human beings, he exhorted. The Bible taught not only brotherly love but also charity and compassion. And activity inspired by these values might be at least as satisfying a palliative as communion with nature, against the fears and uncertainties of the moment—a proposition that worked also no doubt on the psyche of Boole himself.

ON EDUCATION

By the time Boole delivered his lecture "On Education" to an audience at the Lincoln Mechanics' Institute in fall 1848, he had been headmaster of his own school nearby at 3 Pottergate already for some eight years and on school staffs since age sixteen. Not surprisingly, at several points in the lecture he refers to his own practices in the classroom. Rather than providing practical hints as such, however, he attempts here to give a general theory of the relation between different types of instruction and a child's psychological development, taking into account a number of ideas then in circulation, in an especially rich period in the history of pedagogy.

Just when he was writing, demographic, social, and political changes posed new challenges to educators. Calls for electoral reform went accompanied by calls for better and more widespread schooling. Exactly what might be taught was a matter of intense debate. How to deliver the literacy skills necessary for a broader franchise while inculcating the moral values necessary for civic peace? How to apply, on a massive scale, Enlightenment notions about the value of the individual child and normativity of nature?[118] Disagreement about the central issues of instructional content was one of many reasons for the failure in Britain of numerous proposals to set up a state system such as the one implemented in Prussia since the previous century or even a general system of teacher education such as in the Commonwealth of Massachusetts.[119] Religious and secular leaders alike, and their special constituencies, all had their say.

In the very year when Boole spoke, an influential pamphlet announced, "A specter is haunting Europe: the specter of Communism."[120] The old social order was just then breaking at the seams. Marx and Engels spread the gospel of freedom through class struggle. That any impulses from the

1848 revolts throughout the continent might cross the sea to Britain was in the interests of the propertied classes to prevent at all costs.[121] But also in Britain, impatience with the slow pace of legislation and fast pace of worker abasement grew more and more widespread. Boole was keenly aware of the social costs of misery, although he did not put a political label on it. In concert with the social reformers of the time, he noted the close tie between misery and immorality, but rather than blaming the victim, he blamed the cause. "Generally when the condition of childhood has been one of suffering, especially if that suffering has been caused by unkindness, the injurious effects of this violation of Nature's order will be felt in after life."

Ever the egalitarian, at least in matters regarding the distribution of abilities within society, he believed that any youth with a good brain might reach the heights of knowledge through study and application. Considering the extensive use of child labor at the time, this was a highly optimistic view. Nor, in Boole's mind, were social circumstances a necessary determinant of success; other factors were more important. "Whatever takes place within the little world of a child's experience, tends to form his character for good or for evil," he insisted, not "the outward distinctions of birth and fortune." Rather than entering into a polemic regarding the theories of aristocracy that had taken such a beating in the previous century, he noted, "I assume that in their sober convictions this is felt by all to be true." Any claims to the contrary, he implied, violated common sense.

Another reason for the widespread resistance to a national system of education came from deeply held beliefs, articulated in the previous century by Smith, about the dangers of too much government interference in matters best left to the private sphere. Here Boole may have found himself in agreement, although his stated preference for homeschooling over other systems had deeper roots. For him the issue was not so much one of keeping the government out but instead of keeping the family in the schoolchild's life due to the natural love that should prevail among parents and their children. "It is therefore not nearly the dictate of a present benevolence but also of a prospective regard to the future that in all arrangements for the education of the young, their happiness should be consulted. For this reason home education is, under circumstances otherwise the same, preferable to education apart from the kindly influence of the domestic hearth."

That the happiness of children ought to be a matter of consultation in any regard was a point of educational theory that placed Boole in a line of pedagogical thinking that extended from Rousseau to Friedrich Froebel and beyond.[122] Said Boole, "The child is concerned because the happiness and usefulness and virtue of his future life depend very much upon the way in which his years of preparation are spent." A child-centered approach raised such questions as, What kinds of feelings, what kinds of emotions, might be in play in the formation of a young mind? How to awaken the natural tendency to virtue? "It loves its mother before it knows the name of love—and if well-trained and nurtured, it exercises a certain restraint and self discipline over the rising of its little passions and troubles before it learns to attach to that exercise a distinct idea or is acquainted with its importance in the entire conduct of life."

The developmental perspective was as close to Rousseau as it was distant from Dugald Stewart and the elder Mill. Rousseau set out his three-stage theory of infancy, childhood, and adolescence in *Emile: ou, de l'Education*(1762). Nearly a century later, John Stuart Mill revealed the slow change in prevailing ideas in Britain when he famously noted, with some misgiving, that in his own case the instruction provided by his father, James, more focused on acts and ideas than on passions, "was in itself much more fitted for training me to know than to do." In an unpublished version of his *Autobiography*, he revealed that this instruction took no account of the differences in capacity and understanding between a young child and an adolescent, or between a child and an adult, such that he emerged with temporarily impaired competence in aspects associated with such simple tasks as dressing and speaking that children normally acquired at an early age.[123]

In common with some educational theories then in vogue, Boole advocated learning by experience and practice rather than by rote. As he pointed out, "The order of Nature as manifested both in the discovery and the acquisition of knowledge is an ascending and never a descending order." A natural mode of instruction must therefore proceed from the particular to the general, from the concrete to the abstract, and not vice versa. Objects, he insisted, must come before ideas. Thus, in the first verbal utterances, "the child acquaints itself with things before it learns the names of things—it uses its senses before it applies its intellect." By the same token, the matter of a subject was acquired before the rules, such that, for instance, the child "does not begin with the rules of language but with the practice of language—and it is so also in the formation of the moral habitudes."

The inductive method was a feature of English educational thought long before it became associated with the name of John Dewey or with the so-called constructivist theories, although in Boole's view schools of philosophy made more of the supposed dichotomy between empiricists and rationalists than a closer comparison would tend to support.[124] "Some maintain that the office of experience ... is to provide the concrete material from which the mind by its own power of abstraction separates that element which is common to every instance," he explained.[125] Epitomizing the views of Thomas Chalmers, he continued, "Others on the contrary believe that our experience merely serves as the occasion upon which we become conscious of the fundamental truths of our own moral being."[126] Since the logic of ideas went unaffected, from an educator's standpoint, "whichsoever of these views we embrace, the order of the process we are considering remains the same." Boole actually sided more with those, such as the Ulster-born philosopher Francis Hutcheson, who sustained the existence of a particular moral sense, although his precept regarding the lasting effect on later development of even the most seemingly insignificant childhood impressions echoed the sentiments of Locke.[127] In Locke's view, "a foolish maid" might convince a child that "goblins and sprites" are associated with the darkness, such that "darkness shall ever afterwards bring with it those frightful ideas, and they shall be so joined, that he can no more bear the one than the other."[128] The possible social consequences of mass-produced error seemed as urgent a matter in Boole's time as it had been in Locke's. An inductive approach to education offered a way out. One might wonder, one and a half centuries later, what now?

Among the most influential inductive approaches to pedagogy in Boole's day was that conceived by Pestalozzi in Switzerland and propagated throughout Europe and the United States. Nature, not just the reasoning soul, would now be the guide and measure, and children were its creatures. Moreover, the nature of children was to be taken into account in any theory of instruction. "Force not the faculties of children into the remote paths of knowledge, until they have gained strength by exercise on things that are near them." Other statements by Pestalozzi presaged later notions regarding the natural progression of ideas through the mental faculties. "There is in nature an order and march of development. If you disturb or interfere with it, you mar the peace and harmony of the mind. And this you do, if, before you have formed the mind by the progressive knowledge of the realities of life, you fling it into the labyrinth of words, and make them the basis of development."[129] Boole made such concepts the basis of his pedagogical practice in and around Lincoln at roughly the same time as Bronson Alcott did at the Temple School in Boston.[130]

Pestalozzi inveighed against notions of pedagogy then current among the greater portion of teachers, whose methods reflected an outdated training in regard to the mental and moral development of children. "The artificial march of the ordinary school, anticipating the order of nature, which proceeds without anxiety and without haste, inverts this order by placing words first and thus secures a deceitful appearance of success at the expense of natural and safe development."[131] Transcribing dictated texts along with reading and repeating to prepare for interrogations: this was the lot of the typical schoolchild. Once the desire for knowledge was thus thoroughly killed, whence would arise the teachers and thinkers of the future? Such sentiments are echoed by Boole.

Boole proposed his method based on the spontaneous interaction between teacher, pupil, and subject matter. The lecture for the Mechanics' Institute showed how this would work in each major subject. In keeping with the basically humanist approach adopted here, he started with geography and history. In the former, lessons would not begin with the recitation of so-called facts but instead with the introduction of visual evidence in the form of maps. Subjects like arithmetic and geometry should be introduced by way of the things to be counted or shapes to be described, rather than by laws to be memorized. In languages, the pupil's attention was to be directed to the object first and then the rule. English, especially, was to be taught with usefulness in mind. Let pupils be encouraged to write as much as they could whenever they could, rather than filling their minds with grammatical cavils. Boole admitted, "If I have met with any success in the prosecution of literature and science, I am bound to attribute it in a great measure to the habit of writing out, early acquired and perseveringly practiced."

Boole's linguistic competence was impressive and wide-ranging, and his expectations of other learners were uncompromising. German he suggested as a sensible modern language choice, especially for those not interested in pursuing Latin and Greek. "That copious language" he regarded as "the key to a no less copious literature." In promoting this particular national idiom he was in good company. Which of the poets and prose stylists might be worth particular attention, he did not say, although some conjectures are possible. Major popularizers of German Romanticism such as Carlyle had long been directing British audiences to Schiller and Goethe.[132] To more limited audiences, Samuel Taylor Coleridge brought a taste for the school of idealist philosophy and Kant, which John Stuart Mill, for instance, had learned German especially in order to pursue.

An insistence on the variety of competencies belonged to the wider discussions that were just then going on about the ideal of liberal learning. In 1835, Whewell published his *Thoughts on the Study of Mathematics as a Part of a Liberal Education*, where he stated the object of such studies was "to develop the whole mental system of man, and thus to bring it into consistency with itself; to make his speculative inferences coincide with his practical convictions; to enable him to render a reason for the belief that is in him."[133] For him, possibly also because his purpose was to improve the distribution of learning within the university, the discipline par excellence for developing a reasoning faculty was mathematics. For Boole instead, at least in this lecture regarding early education, the paradigmatic reasoning science was taken to be language. The significance of this emphasis from the standpoint of his eventual understanding of the interdisciplinary study of logic cannot be ignored.

To at least a few in the audience, the linguistic approach should not have been too surprising. A year before this lecture, Boole published his *Mathematical Analysis of Logic* with the first outline of his symbolic algebra. There he drew the epigraph from Aristotle's *Posterior Analytics* (1:11), stating that "all the sciences share with one another in the use of common principles." He introduced the work with the reflection, "The theory of Logic is … intimately connected with that of Language." Ultimately, he went on, "a successful attempt to express logical propositions by symbols, the laws of whose combinations should be founded upon the laws of the mental processes which they represent, would, so far, be a step toward a philosophical language."[134] Language, logic, and mathematics at this point therefore perfectly aligned. In his lecture, he suggested that "justly considered, the theory of language depends very much upon the laws of the mental faculties of classification, so that instruction in the science of grammar may be made simultaneous with instruction in the science of reasoning."

For most of the lecture, Boole speaks of the disciplines in general terms without going into the specific matter being taught. In one instance alone, that of the teaching of the ancient languages, he makes an observation regarding what topics to choose, or rather what to omit. There was far too much at stake here to leave anything to chance. Like any thorough believer in the doctrine of progress, from the standpoint not only of the disciplines but also of faith and morals, Boole could not let his love of antiquity blind him to the defects, as he saw them, of the pagan past or indeed "the vices of Heathenism." Not only were the ancient Greeks and Romans ignorant of the truths of Christianity. Perhaps not unrelated to this, they were dangerously hedonistic as well. Those well schooled enough to view these writings as products of a less enlightened time and place might be spared the pernicious effects. Those also whose convictions were already well formed might escape damage. "To the pure all things are pure." Yet "there are I suppose few whose classical reading, if of any extent, does not furnish them with passages in ancient authors which, if right minded, they regret ever to have seen" due to being "tainted with a moral pollution which … has no place in a Christian literature." The best modern writers, however steeped in classical lore, have for the most part wisely prevented such aspects from contaminating their works. "The native literature of this age, though in all other respects free beyond that of any former, does not even tolerate by suggestion those things which appear without disguise on the classic page." Let the wise instructor therefore protect the young from such ordure by directing attention only to that which is morally uplifting, allowing them at a later time to choose to look deeper and learn. "Whatever may be the discipline proper to a riper age, the thoughts of the young should be made familiar only with whatsoever things are lovely, whatsoever things are pure."

Here as elsewhere in these lectures, the moral vision went accompanied by a natural theology. Not only did belief in a transcendental power constitute a necessary part of the order of nature, such that Boole, speaking of the pupil, referred to the "catalog of the different objects of his experience," which obviously must include "the sun and stars, trees, plants, furniture and etc." as well as the objects of his belief, such as "God, the soul etc." Moreover, morality and divine precept were one and the same. To Chalmers, the only author cited by name in the text, he attributed an insight regarding "the connection between the great fundamental principles of morals and those of the Christian religion."[135] Indeed, "there is no error more injurious than that which represents religions and true morals as in any degree opposed to each other," Boole asserted. Instead, "the primary laws of morality" may be regarded as "the immediate expression of the character of their Author, a view in which it is necessarily implied that they are the foundations of all religious truth." Here as elsewhere, he declined to go into detail "for obvious reasons" considering the philosophical and doctrinal subtleties that the occasion did not permit.

THE CLAIMS OF SCIENCE

After an application process that extended over several years, in 1849 Boole finally received the nomination as professor of mathematics at Queen's College, Cork. His reputation among the learned was already such as to overcome any doubts about his fitness for the job that might have arisen due to the lack of a university background—as he himself had hopefully suggested in his application to George William Frederick Villiers, fourth earl of Clarendon, then the lord lieutenant of Ireland. He arrived in Cork in October, a month before the ceremonial opening of the new college, and within two years he was named dean of the science faculty. In that capacity, he gave the address that is published here.

The fledgling institution was already embroiled in controversy, as Boole seems to acknowledge toward the end of his opening address: "There may be periods in which the prospects of science, and with them those of human improvement, are sufficiently discouraging. The strong tide of party may set against it. Detraction may assail its friends, misrepresentation sully and distort its beneficent aims." The new college, conceived as a nondenominational alternative to the avowedly Protestant Trinity College in Dublin, managed to attract scorn from both sides of the major confessional divide. Its alleged lukewarmness to the callings of faith only fomented the fierce ideological battle surrounding a recent papal initiative to assign more bishops to Britain. The Catholic synod of Thurles in 1850, declaring it off-limits for clergy and inadvisable for laity, proposed an alternative Catholic University.[136] Boole, who rarely allowed his confessional orientations to enter explicitly into his work, may have found it curiously hospitable, in spite of the personal foibles of a few colleagues.

To be sure, the professor of mathematics was in no way exposed to any of the same perils as was the professor of modern literature, his friend and housemate, an expert on Dante and Milton, named Raymond de Vericour. When the latter's history textbook was placed on the Roman Index due to an alleged antipapal bias, the local Catholics successfully lobbied for the author's dismissal, which later was grudgingly rescinded.[137] As Boole never tired of saying, at least in regard to particular denominations, science ought to be confession blind. Nor was this the age of Galileo, when the truths of the Bible were thought to be best proclaimed by muzzling the mathematician. And whatever might be the controversies of the moment, he assured his audience in the lecture, science was destined to prevail: "Obscured by the mists of prejudice, forgotten amid the strife of parties, she but the more resembles those great luminaries of heaven, which pursue their course undismayed above the rage of tempests, or amid the darkness of eclipse."

The benefits of human reason applied to problems of practical life were there for all to see. Just when he would have been speaking, the Great Exhibition was being held in London at the newly built Crystal Palace, an innovative structure in steel and glass.[138] For the duration of that event, some six million people would come to view the industrial and technological progress of "the workshop of the world," ultimately the result of imperial arrangements and a capital economy—not only in Britain. The *Economist* magazine was impressed. Not only could observers appreciate the eventual benefits of the current system. If only countries would forget their petty differences and join this "pilgrimage of all nations to the shrine of peaceful industry," a happy future was guaranteed. The building and diffusion of new, more powerful machines for transporting passengers and goods, cultivating the land, and producing commodities and all the necessities of life promised to guarantee longevity to European civilization. Indeed, in a slightly playful vein regarding current providentialism, "those who have lived through this wonderful era will lose all regret at not being suffered to witness the yet more wonderful things that are in store for their predecessors, in the bright hope that they will be the produce and the reward of the ingenuity and virtues they have been permitted to behold."[139]

Concerning the exhibition, which he visited in June 1851, Boole's only recorded remark is "it fully answers all my expectations."[140] But if there was any other faith besides Christianity to which he adhered unwaveringly, unquestioningly, fervently, that was a faith in the progress of humankind—especially through the increase in knowledge.[141] Such was the chief theme of his opening address in October of that year. There he articulated the doctrine of progress toward greater enlightenment that his age inherited from the eighteenth-century philosophers. And there he laid out a doctrine of hope in the midst of adversity to inspire the dreams of many. "Every science," he explained, "is … a gradually increasing system of knowledge which, beginning with experience, advances ever onward through successive stages towards that perfection which no science has yet reached, which none perhaps ever will reach." The final vision rose high above the gloom of despair: "The idea becomes clearer and brighter, with every approach that we make."

Whatever may have been the approaches, as adopted by the different disciplines, "science" in Boole's exposition was always one and the same. It was "the joint result of the teachings of experience, and the desires and faculties of the human mind." It normally began with sense experience, submitted to operations "of comparison, of reflection, of reason, and of whatever powers we possess, whereby to perceive relations, and trace through its successive links the chain of cause and effect." As he often repeated (see the lecture "On Education"), "the order of its progress is from particular facts to collective statements, and so on to universal laws." It did not differ between the natural and mathematical sciences, since "in Nature it exhibits to us a system of law enforcing obedience, in the Mind a system of law claiming obedience. Over the one presides Necessity; over the other, the unforced obligations of Reason and the Moral Law." In fact, against the skeptics, he affirmed, "We may be deceived by external appearances, but this source of error apart, the uniformity and universality of the laws of nature is, so far as the range of its just application extends, the most solid foundation of human certainty to which we have yet attained."

This sentiment regarding the unity of science aligned Boole as closely with contemporary conceptions as it distanced him from some more recent ones. James B. Conant, for instance, writing in the early years after World War II, considered that practitioners in Boole's time were still groping in the dark as far as the scientific method was concerned. And he carefully distinguished the experimental approach of the natural sciences, as perfected in a later age, from the approach prevalent in other disciplines, decrying the "extreme position which has been maintained with some insistence for the last seventy-five years" that tended "to equate the scientific method with all relatively impartial and rational inquiries."[142] Nothing, he insisted, could be further from the truth.

In Boole's account, not all the discipline-specific modes of inquiry contributed to knowledge in the same way or at the same rate. To the extent that they relied less on laws of nature, they might be more prone to error. For example, "I shall not here pause to dwell upon the social and economical sciences, which regard men, not as individuals, but as members of a community, and sharers of a public interest, and which are based upon the consideration of prevailing motives, rather than the requirements of an ideal standard of conduct." Here human nature insinuated a measure of uncertainty. "As men cannot be divested of their individuality, such sciences do not profess to attain the formal strictness of those which have been already considered." He went on, possibly referring to entomology or ornithology, to observe, "I pass over in like manner some other departments of knowledge, which, depending chiefly upon classification, may be regarded as the precursors of science, rather than science."

Such distinctions did not affect the overall theme of progress, which Boole inherited from his intellectual forebears. Only a half-century before, in his *Sketch for a Historical Picture of the Human Mind*, the Marquis de Condorcet had described the tenth or last stage of social and mental development, corresponding to modern times. "Nature has fixed no limits to our hopes," he proclaimed, for infinite perfectibility. There would be better education, more tolerance, a more rational use of economic resources, improved public health, and consequent longevity. Humans would improve, and the improvement would be permanent. "In this persuasion," the French probability theorist went on, we "find the true delight of virtue." Awaiting the guillotine in a Jacobin jail, such contemplation afforded him "the asylum to which he retires, and to which the memory of his persecutors cannot follow him."[143] He carried his optimism with him to the grave.

For Boole, writing under somewhat safer circumstances than Condorcet, significant practical improvements in the way most people lived their lives would accompany the mental improvements. He could "anticipate a period when the physical evils which afflict our present state shall exist no longer, or exist in such measure only as is inseparable from a condition of mortality." Likewise he could "anticipate that in this happy state of things to come, relieved from the oppressive bondage of physical wants, man shall be at liberty to accomplish, and actually shall accomplish, the higher ends of his being." Rising in this joyous reverie to the limits of the imagination, he went on, suggesting "that while the earth shall shine with more than its pristine beauty, the human family shall not only be clothed with the fair assemblage of the moral virtues, but shall add to them that crown and safeguard of knowledge which has been won from the hard experience of ages of error and suffering."

Perhaps with a regard to the laboring poor of his adopted home, whose abject condition he noted to his correspondents almost on arrival in Ireland, in this lecture too he foresaw a better future "when painful toil shall have been replaced by the appliances of mechanism." And in view of the biological calamity at his doorstep, he foresaw "when the most prolific sources of disease, as crowded cities, undrained swamps, pernicious indulgences, shall have disappeared before a more enlightened study of the conditions of health, and a truer appreciation of the ends of life." Penetrating to the deeper avoidable and unavoidable causes of current troubles, without pointing any fingers too directly, he envisioned a time "when the excessive inequalities of wealth, and the miseries which they entail, shall have yielded to a better moral or social economy; and when the effects of those casualties which prudence cannot avert, as earthquakes, tempests, unfriendly seasons, shall either be reduced to a minimum of amount, or shall be so distributed as to fall with the least oppressive weight upon the community at large."

Boole's account of a science-led future was of course far less systematic than the one just then being developed by Comte, founder of the doctrine of positivism. Having reduced Condorcet's ten-stage theory to three, beginning with the "theological" and "metaphysical," Comte welcomed the advent of the third "positive" stage in which "the explanation of facts, reduced to its real terms, is henceforth nothing but the connection established between the diverse phenomena and some general facts, which the progress of science tends more and more to diminish in number." Now, "with the human spirit having established celestial physics, terrestrial physics both mechanical and chemical, and organic physics both vegetable and animal, there remains only to complete this system of sciences by establishing social physics," which he termed "the greatest and most pressing need of our intelligence."[144]

Mill was so impressed with Comte's idea that he published an entire book on it in 1865, but already in 1843, in book 6 of his *Logic*, he endorsed the historical vision. "By its aid we may hereafter succeed not only in looking far forward into the future history of the human race, but in determining what artificial means may be used ... to accelerate the natural progress." Like Comte, he held out hope for the application of a specific and distinctive scientific approach to those areas most concerned with human flourishing, especially in the realm of human interaction. What worked in physics, though, did not necessarily work in society. "Whenever the nature of the subject permits the reasoning process to be without danger carried on mechanically, the language should be constructed on as mechanical principles as possible; while in the contrary case it should be so constructed, that there shall be the greatest possible obstacle to a mere mechanical use of it."[145] Boole went so far in agreeing with this last proposition that he quoted it in his *Mathematical Analysis of Logic* in 1847.

For Boole, what joined the various knowledge-making enterprises was not a single method but instead a remarkable coincidence in regard to the material on which they worked. Kant considered that this coincidence derived from nothing other than the common origin of the method and material. Both were part and parcel of the mind's operation. The external world, such as only humans can apprehend it, was conjured up by the mind out of sense impressions—the same mind, indeed, whose operations produced conclusions about the perceived phenomena. Boole agreed with Kant that there was also "a real phenomenon in the mind ... which distinguishes it from the system of external Nature."[146] Chalmers had developed a similar line of thought in his treatise *On the Power, Wisdom, and Goodness of God, as Manifested in the Adaptation of External Nature to the Moral and Intellectual Constitution of Man.*[147]

Yet he elaborated on the suggestion that the human mind is aptly designed to achieve truth regarding precisely those questions to which it directs its attention.[148] "If we are sensible of the existence of faculties and powers whose province it is to detect order amid apparent diversity, to discover the indications of cause amid the seeming results of accident, those faculties do not exist in vain." There was a higher purpose at work in the relation between the observer and observed. "The mind of man is placed amid a scene which can afford to all its powers their appropriate exercise." How not to see the work of providence here? "There is thus a correspondency between the powers of the human understanding and the outward scenes and circumstances which press upon its regard. In this agreement alone is Science made possible to us. The native powers of the mind, cast abroad amid a world of mere chance and disorder, could never have realised the conception of law." The mission of the man of knowledge was to apply those powers to those objects of inquiry for the benefit of humankind and introduce others to the rewards of right reasoning.

Indeed, anyone who doubted the rewards of right reasoning had only to look around at the disastrous consequences of error and misdirected conviction. Urban squalor and disease, wherever Boole had experienced it, furnished a ready repertory of examples. "Careful inquiries assure us that there is a real connection of cause and effect between an undrained, uncleansed condition of our towns, and the prevalence of fever and a general high mortality. I suppose that there are few conclusions better established than this." In particular, the Irish cholera epidemic of three years before had notoriously killed off a good number of famine survivors, and there were few signs that the same could not happen again.[149] This was the "fearful confirmation" to which he referred, "when some epidemic disease, making head against all the resources of medical art, emerges from the dark lane or the noisome alley, and sweeps away the rich and the poor in one indiscriminate destruction." If the immediate prospects were not good, this was to be attributed not to the conditions but rather the (sometimes self-interested) refusal to pay attention to the biology involved. "Men are, however, for the most part, so reluctant to admit the reality of that which they do not see with their eyes, that this teaching of science, and this confirmation of experience, are sometimes alike void of effect." By this time, the microorganic generation of morbidity was accepted wisdom among the expert community, but the uninstructed among the general population and even in responsible offices might easily be fooled. "They cannot perceive with their bodily senses the connection between impure air and disease, and they refuse to believe in invisible laws; or, if they acknowledge them in words, they do not give them any hearty assent. And so the scene of desolation is renewed from year to year."

Nonetheless, in spite of the stops and starts, in spite of the refusal of some and weakness of others, in spite of the propensity of many to sacrifice long-term benefits for short-term gains, the sum of human experience suggested a natural law of progress, and the laws of nature, in this realm as in every other, were ineluctable. They were involved, he implied without articulating, and this conviction again belonged to other votaries of natural theology, with the providential plan of the Author of all.

THE SOCIAL ASPECT OF INTELLECTUAL CULTURE

News of the Crimean War reached Cork in 1853, with the *Irish Examiner* echoing similar coverage from the London *Times* and other papers. To many, the danger may have seemed far-off. Still, young men actually left the following year from Kingstown harbor near Dublin (Dún Laoghaire) to go and fight with the British Army against Russian expansion into the lands of the declining Ottoman Empire, alongside the French and Sardinian allies. Historiography at least since A. J. P. Taylor has turned attention away from the war's proverbial uselessness and toward the long-term consequences.[150] At the time, eventual victory on the allied side did nothing to erase an impression colored by murky objectives, a brazen power struggle, brutal combat, and unprecedented press coverage that had tongues wagging wherever there was information. The futile Charge of the Light Brigade, reported in November 1854 and captured in widely circulated verse weeks later by poet laureate Alfred Lord Tennyson, seemed to epitomize the typical mix of hubris, carelessness, and courage.

Boole, rather than referring to that poem, in his address on May 29, 1855, evoked Milton's war-singed sonnet 21 instead, as a reminder of how the average citizen may be distracted from the good things of life by thoughts regarding the geopolitical concerns of the day—symbolized as "what the Swede intends, and what the French."[151]

Fit for accommodating such reflective recreation, a new multipurpose assembly hall dubbed the Cork Athenaeum had recently been built in the center of town at the Lee River's edge by Richard Rolt Brash based on a design by Sir John Benson, financed by the Royal Cork Institution, an educational foundation, through private contributions. Later in the decade, Dickens would stop here on a reading tour throughout the British Isles.[152] Although the space was already in use since January 1855, the official inauguration by George Frederick Howard, the earl of Carlisle and lord lieutenant of Ireland, did not take place until May 23. By that time the Cork Cuvierian Society, a gathering of learned amateurs, knowledge professionals, and local luminaries, had nearly completed preparations for an exhibition of objects—artistic, industrial, natural, and historical—calculated to excite the interest of the curious. As president of the society, Boole explained the endeavor.

Public-spirited associationism was still alive and well in Cork, in spite of many setbacks. Some of the same causes that had stimulated the Mechanics' Institutes in England were in operation here, except that the trend in electoral reform worked in the opposite direction: rather than tending to an expansion of the franchise, there was a move to restriction on economic grounds in order to block the widespread, less prosperous Catholic population, especially after the Catholic Relief Act in 1829 allowed Catholic membership in Parliament.[153] Rather than preparing a broader electorate, here the inspiration behind the institutes came from hopes for more general social, political, and economic renewal and self-improvement. The mission to counter industrial backwardness brought temperance enthusiasts together with Catholics and local opponents of the 1800 Act of Union joining Ireland to a so-called United Kingdom, and between the 1820s and 1860s, institutes were founded in Cork and no fewer than thirty other Irish locations, where mixed audiences listened to speeches heralding the latest in technology and research of all sorts.[154] The landscape of public science (or public knowledge) was further nuanced by a variety of other institutions, and apart from the Royal

Cork Institution, there was also the Cork Literary and Scientific Society, founded in 1820 and still ongoing as I write.

The Cuvierian Society's expanding role at this time appears to have been a function of the gradual decline of the Royal Cork Institution from which it sprang, combined with a series of intelligent leaders favored by a willing audience. The French connection was evident in more than name only, considering that the first president, James Roche, a prominent banker and patron of the parent institution, had been educated in France and imprisoned during the French Revolution, just when Georges Cuvier was beginning to assume a prominent position in French science.[155] Five years after the founding, in 1840, members' contacts caused the gift of a portrait bust and selection of publications to be donated by the widow of the great naturalist. This did not mean that intellectual affinity with the theories was in any way a prerequisite for members, even if the new approach to the notions of life and biological structure could seem like a reasonable answer to growing dissatisfaction with species transformism as propounded by Jean-Baptiste Lamarck, also considering the theological implications.[156] What kept attention focused on the society's activities was less a particular intellectual doctrine than a continuously interesting schedule of events and demonstrations. Publication of the *Contributions toward a Fauna and Flora of the County of Cork* in 1845, authored by J. R Harvey, J. D. Humphreys, and T. Power, signified the serious ambitions.[157] So did the participation of Joseph Ellison Portlock, a specialist involved in the Ordnance Survey in Ireland, along with present and future Queen's College staff, such as William K. Sullivan, a beneficiary of chemical instruction in Germany in the age of Justus von Liebig and later president of the college.

The minute book and published records of the Cuvierian Society in Boole's time testify to an impressive range of lectures and demonstrations across the arts and sciences. In March 1855, an electric induction coil of the sort devised by Heinrich Ruhmkorff was introduced along with associated mechanisms to produce curious effects in heat and color.[158] In the same period, a demonstration of the stereoscope showed the wonders of three-dimensional photography.[159] A stuffed kiwi brought on a discussion of the nest and habitat where the specimen had been found along the Hokianga River in New Zealand.[160] A report on experiments and observations weighed various potential causes of the potato blight, ranging from fungus to insect related, and compared results with those obtained by the London Horticultural Society.[161] An account of the destruction and dismemberment of Irish manuscripts for commercial purposes was a reminder of the fragility of cultural heritage.[162] Boole himself spoke on one occasion about the scholastic philosopher Robert Grosseteste, bishop of Lincoln (d. 1253), best known for a theory of vision. On another occasion, he explained representations of the cosmos in current theories and a Ptolemaic model from the fourteenth century.[163] On yet another occasion, he recounted the odd story of a Limerick-born mathematical eccentric named John Walsh who claimed that calculus, invented by Newton and Leibniz, was a colossal error. In a discussion of decimal coins, weights, and measures, Boole suggested the possibility of introducing such a system into the United Kingdom on the French model.[164]

For a major event with an international outlook, featuring local resources and capacities, there was a powerful precedent—not only at the Crystal Palace in London. Just three years before Boole gave his "Social Aspect of Intellectual Culture" lecture, the first Irish Industrial Exhibition opened in Cork to celebrate a "movement"—in the words of Cork mayor John Francis Maguire—that could scarcely fail to raise the country from its knees, considering the evident "determination" to foster economic development "by every practical and legitimate means at their disposal." Stands featured agricultural implements and industrial machines as well as products like butter, ale, porter, paper, linens, poplins, velvets, feathers, music strings, boots, shoes, clocks, watches, and artificial teeth, along with three "new branches"—namely, fancy biscuits, confectionary, and pearl barley. The list went on, ranging from carpets to drugs, lithography to chandeliers, and a special section devoted to the Portlaw Cotton Factory. A "female industrial movement" was touted, with news from schools and shirt factories. Painting and sculpture were featured, with works by James Barry, William Gorman Wills, John Hogan, Patrick MacDowell, and James Heffernan. "Curiosities of natural history" received due attention as well as Irish antiquities, including Ogham stones, brooches, harps, and a Saint Patrick's bell. Altogether, some 140,000 visitors were counted over the duration of the event. Adding a wild surmise about the root of any present troubles, accompanied by a typical note of class condescension, the mayor professed his "earnest hope" that the example might be imitated "throughout the length and breadth of this lovely and fertile island, whose beauty has been so sadly marred, and whose abundance has been so fatally neutralised, through the mad dissensions of its highly gifted but too impulsive children."[165] Months later his plea was answered in part, and William Dargan, a railway magnate from Carlow, personally sponsored an exhibition of industry and arts in Dublin, to which the most distinguished in attendance, with all the obvious implications, were none other than Queen Victoria and Prince Albert.[166]

There was no time to lose. As opportunities on the island seemed to disappear along with the supply of willing takers, the drastic loss of population due to death and disease between 1845 and 1850 was compounded by massive emigration. By 1855, two million had left, and no amount of enthusiastic boosterism would bring them back or cancel the economic reality filling the media space with advertisements for passage out of Ireland.[167] We are almost inclined to wonder whether the dizzying swirl of numbers, none of them good, starting with the terrifying census report of 1851, was what Boole was thinking about when he quoted Milton, "Let Euclid rest, and Archimedes pause."[168] One certainty is inescapable, amid the continuing debates regarding the causes and extent of the postfamine conjuncture. If those who remained were to continue streaming into the cities from the troubled countryside, like their fellow workers all over Europe in this period of massive and massively uneven urbanization and industrialization, they had to have something to do.[169]

The Cuvierian Society's exhibition at the Cork Athenaeum involved a selection of items from various collections systematically organized to reflect the various "departments" into which the society's labors were subdivided—with the exception that "Statistics and Political Economy" were omitted, perhaps for reasons having to do with the chiefly upbeat tone of the event.[170] Each exhibit came accompanied by a speaker to explain the items and their importance. "Antiquities" included the "Book of Lismore," a precious fifteenth-century manuscript of Irish writings on loan from its current owner, Thomas Hewitt.[171] Ancient Irish weapons shared space in this section with ancient coins and medals, drawings of sites including the Ardfert Cathedral ruins in County Kerry, and some Greek and Etruscan pottery.[172] "Art" included the Wedgewood reproduction of the Portland Vase as well as various items of portraiture, including representations on Sèvres porcelain and in stone.[173] "Natural History" included an array of drawings from microscopic observations, a variety of stuffed specimens such as a tiger "in a couchant attitude," a full-grown otter, a leopard, an opossum, various birds such as a peacock, and the detached horns of a buffalo.[174] There were live mosses and ferns, some from Oceania, and a sample of the American waterweed that had been introduced recently into the United Kingdom with not exactly beneficial results. The "Physical and Experimental Sciences" involved the usual crowd-pleasing array of electrical instruments for demonstrating "Galvanism, Magnetism, Electro-Magnetism and Magneto and Thermo-Electricity." Various pneumatic devices were on view, and likewise hydraulic ones, including a press set up to break three-quarter-inch iron bars with seeming facility (and possibly much noise) to "no small astonishment" of everyone who watched.[175]

Daily attendance for each of the three days of "conversazioni," as the society termed these meetings by the usage just then in vogue, amounted to some fifteen hundred people—of whom several hundred were trucked in for the occasion from nearby Ballincollig by their employer Thomas Tobin, president of the Cork Athenaeum and owner of a gunpowder mill.[176] Boole recorded the event as a resounding success for the Cuvierian Society, which clearly fulfilled its current mandate for extending its range to comprise not just the study of natural history but also everything else besides—"and, at present, no limits are recognised but those which good sense prescribes." Moreover, the success was for Cork and the country, as the items on view and relevant explanations were such as to "exemplify the progress of the peaceful arts" as well as "open up to us the life of other times," and if this proved impractical, at least to refer to matters "in some way connected with the interests of humanity." Again, in subtle reference to the context of a troubled Europe, he recommended "enjoyment of the presence of those objects which, as they are the fruit, so they are the symbols, of peace and good will among men." Finally, the success was for individual self-improvement through education, instruction, and the right use of leisure.

Boole's doctrine regarding activity and repose, in this lecture, echoed similar sentiments expressed by Kane in *The Industrial Resources of Ireland*—with some key differences. In Kane's view, the gentler arts were themselves a form of recreation, not part of the process of invention per se. One could contrast them to the life of action, as a form of repose, suited to full-time engagement only by those whose existence demanded no real productivity—"those, who, independent in fortune, and devoted rather to ease than enterprise, wish to dream through an existence which offers to them but roses they did not plant, to seek in the literature of past ages, an elegant and innocent occupation." Even those who worked for a living would likely pay attention to such subjects only to rest the mind. "To all classes the literature of present and of past ages, of our own and of foreign countries, presents a relief from the weary continuity of action, which industrial progress requires." Insights there might be, but not about essential things. Indeed, "to the man of business, there can be no enjoyment greater than to transport himself from the anxieties of the desk or factory, to communion with the best lessons, which human intelligence has handed down, to obtain within a few volumes, the records of the greatest deeds, the noblest struggles, and the holiest thoughts which have been allowed to man." Yet "this is not his business."[177]

For Boole, the pursuit of knowledge was a form of activity in itself, if taken seriously, and the study of great intellects at work in the pursuit of knowledge was the necessary concomitant to any consideration of how ideas were born, grew, and could be applied to human flourishing. "Now, this connexion between intellectual discovery and the progressive history of our race, gives to every stage of the former a deep human interest." Synthesis and pattern recognition allowed every field of study to apprehend the inner essence and causes of things, otherwise seemingly mysterious. "Each new revelation, whether of the laws of the physical universe, of the principles of art, or of the great truths of morals and of politics, is a step not only in the progress of knowledge, but also in the history of our species." History and literature might be pursued idly— and in his final paragraph, Boole cast aspersions on a pointless antiquarianism. But they were not idle pursuits per se. "Could we trace back our intellectual pedigree, if you will permit me to use such an expression, we should find ourselves connected by that noblest of all lines of descent, with every nation and kindred of men that has occupied a place in history, and with many others, of whose names and deeds no record survives." Such reflections could be a source of pride, inspiration to accomplishments, and lesson in method. For instance, interpretation of the ancient monuments might show the cumulative effect of humanity's effort to communicate meaning and record its past. "We should see the picture writing, most probably of some forgotten Asiatic tribe, passing through successive stages, analogous to those which are still preserved in the monuments of Egypt, until among the Phoenician people it gave birth to our present system of letters." Careful observation would reveal the font whence civilization sprang, and the answer was a structure, not a place. "We should behold the first principles of our science, and much more than the first principles of our literature and philosophy, emerging into light among those isles of Greece, which seem to have been the chosen home of freedom and of genius in the ancient world."

Freedom and genius in Boole's view went hand in hand, and a good society ought to be capable of accommodating both.

BOOLE AND BEYOND

Such, then, were some of the deepest thoughts and thought processes of a major nineteenth-century thinker, inasmuch as we are able to discern them, and such are the main ideas within the whole corpus of material presented in this book. Much remains to be done. I have hardly mentioned the ideas for which he has become best known. Why did he think as he did? So far I have implied that he drew conclusions on the basis of what he considered to be the best evidence available to him, evaluated according to the methods he considered appropriate to the particular type of knowledge in question. Often he relied on the authority of productions that commanded his respect because of the reputations of their writers or quality of their argumentation. Everywhere we see the logical mind at work. When he formed his convictions regarding a particular matter—the importance of pluralism, the meaning of mythology, the role of education, and the value of freedom—he drew on a vast store of influences and experiences rooted in the world of early Victorian England. Some versions of the history of ideas would call on us to assess his conclusions, or explain his results, on the basis of the strength or weakness of his evidence or reasoning, from the standpoint of the time and even another time—such as our own. In my approach to the vexed question "the social construction of what?" I have tried to make a case for viewing Boole not as a figment of our imaginations but as a creature of a particular context.[178] I have wondered what aspects of that context are most important. What role did the institutions play? What about the structure of knowledge? I offer no definitive answers here, but some conjectures are possible.

In an age of narrow specialization, an intellectual range such as Boole's may seem outrageous. His was no such age. To reinforce this point without suggesting the equally broad sweep of a contemporary intellectual like Mill, one might cross the English Channel and review a Prussian case such as Rudolf Virchow. Younger than Boole by six years, he completed gymnasium in Pomerania in 1839 with a theological thesis titled "A Life Full of Work and Toil Is Not a Burden but a Benediction." He followed his own precept, and after he abandoned theology, his contributions to cell structure, pathology, parasitology, veterinary medicine, public health, anthropology, and paleopathology earned him worldwide acclaim and a number of named structures. Unlike Boole, he managed to outlive the early nineteenth century of polymaths and generalists, and flourish in the emerging university world of discipline-specific departments.[179]

The nature of knowledge in Boole's time was reflected in the institutions. At Queen's College, Cork, subjects were organized in a fashion to reflect time-honored traditions rather than new disciplinary categories.[180] The arts faculty was divided into science and literature, of which literature comprised the following areas, with one professor each: Greek, Latin, modern languages, Celtic languages, and a combined unit called "History and English Literature." Thus, history was not a discipline in itself, nor was English, nor, say, German. On the science side were mathematics, chemistry, mineralogy and geology, civil engineering, agriculture, and three curious relics of an earlier worldview—that is, "Logic and Metaphysics," natural history, and natural philosophy. When Boole, in his essay "On Education," said, "I believe that there are very few studies so remote from each other and so unconnected that they may not in some way be made to contribute to their common furtherance," he was simply stating an accepted fact reflected in the institutions.

To get a flavor of the early nineteenth-century episteme, no lectures from Boole's colleagues at UCC as such remain. We may infer the contents of "natural philosophy" there and elsewhere from manuals popular at the time. The definition in Quetelet's *Summary of a Course of Natural Philosophy*, translated in 1832, could have been written in Newton's day: "The objects of natural philosophy are, the observation of natural appearances, the experimental investigation of the manner in which they take place, and the determination of the forces by which they are produced."[181] Thomas Young's course on the same subject, updated by Philip Kelland in 1845 and present in the college library in Boole's day, divided the material into mechanics, hydrodynamics, and physics. Hydrodynamics included sound and music, optics, and ocular physiology. Physics included electricity. As late as 1867, Lord Kelvin and Peter Guthrie Tait gave the title *Treatise on Natural Philosophy* to a work that essentially dealt with the role of energy in physics.

Worth mentioning here is that "natural philosophy," considered in its more general aspect, still doubled as a definition for "science" as a whole, and would continue to do so until the modern term came into general use.[182] Arguably, the modern institution encompassing a community dedicated to the scientific pursuit of organized knowledge by definite methods according to agreed-on standards of evaluation, research, and conduct belongs to the last third of the nineteenth century and not before, and the interpretation (still hotly contested) that attributes the strongest form of this definition exclusively to the natural sciences dates from that time too. Similar things were going on in other languages. A shift from the Romantic understanding of *Naturphilosophie*, shared by Goethe and Friedrich Wilhelm Joseph von Schelling, to a more mechanistic version of natural knowledge, facilitated a migration to the concept of *Naturwissenschaft* and something like the institution implied in the English term science. But we are already far beyond the scope of a book on Boole.

Natural history, another element of the Queen's University curriculum, comprised, according to one of the broader interpretations, "The natural history of animals, with human and general animal physiology, botany, vegetable physiology and geology."[183] Another author specifies, "The course of Natural History then, primarily and essentially, according to the universally received meaning attached to the term, involves Zoology." The author adds, "Zoology as a science, is one of wide-embracing comprehensiveness; the animal creation is its subject, but the animal creation viewed from a thousand varying points."[184] The main work to be reckoned with, if only in refutation and animadversion, was still Georges Buffon's massive *Histoire Naturelle, générale et particulière, avec la description du Cabinet du Roi*, whose thirty-six volumes appeared between 1749 and 1804, available in English in various translations, with the latest (from Boole's perspective) in 1822.

Mathematics was always something of an anomaly. Having entered the first universities largely via medicine due to antique notions about the influence of the planets on the human body, it retained a connection to practical fields like astronomy due to the demands of spherical trigonometry.[185] Mathematical applications in ballistics and civil engineering ensured extramural engagement. By way of physics and mechanics, it had opened the door through which science subjects other than medicine and physiology had entered curricula originally designed for the professional qualification of doctors, lawyers, and theologians. Considering the wide remit, a writer for the *Monthly Magazine or British Register* in 1821 had to insist, against the prevailing current, "a mathematician is not necessarily a philosopher."[186]

A pattern of amateur engagement in British science comforted the generalist without disparaging expertise.[187] Although Boole was in good company, times were rapidly changing. Already in Germany, academies and private laboratories were losing ground as centers of research productivity to publicly funded university institutes.[188] A generation of practitioners with no advanced degrees would eventually be replaced by PhDs. Alexander von Humboldt pioneered the idea that teaching should be guided by the latest research, and research belonged in the universities—a concept honored thus far only by a few. At the dawn of the twentieth century, the transformation had reached such a point that Max Weber lamented the advancing hegemony of "specialists without spirit."[189] To what degree the particular organization of knowledge in early Victorian England contributed to Boole's best-known works is still an open question.

If Boole never developed his side interests into a veritable expertise this was not likely due to a lack of ability. Had he more time away from his practical pursuits to devote to other topics besides mathematics his output might well have been different. Indeed, what works he may have had under hand at the time of his death is by no means clear. The manuscripts as first deposited in the Royal Society are nowhere fully described, and they arrived in a highly disorganized state in 1873 according to the society's historian Marie Boas Hall: "[The] papers ... were donated by his widow in a somewhat informal manner: she left them in a bag with the Burlington House porter, so that it was not clear to the family whether they were donated or merely left on deposit, nor is it clear whether Mrs. Boole sent the papers she found the next year to the Society or not." Matters regarding the papers continued to be pursued in this fashion, such that "in 1889 [Boole's] daughter was to write to Mrs. Rix [the archivist], whom she knew, explaining that she would like to borrow the papers for the use of friends who were planning to republish Boole's *Laws of Thought* (1854); she was given permission to do so, but the Society had great trouble in getting them back and did not succeed for nearly ten years."[190]

The contents of the nine (plus two additional) boxes at the Royal Society is highly miscellaneous. Apart from copies of Des MacHale's book on Boole and Souleymane Bachir Diagne's *Boole, L'Oiseau de Nuit*, neatly bundled away in an archival folder, we find draft autograph pages, in MSS 782, box 2, for a writing "On Belief and Its Relation to Understanding," another on "The Origin of Evil," and reading notes on August Neander's *Life of Jesus Christ*. Box 9 contains what appear to be reading notes, transcribed and typed up at some later date (unclear by whom) on Nassau William Senior's *Political Economy* and Joseph Butler's *Sermons*. In an additional box 1, there are notes to Ralph Wardlaw's *Christian Ethics*. The best description so far is in the anthology by Grattan-Guinness and Gérard Bornet.[191]

University College Cork acquired its share by purchase only in 1983, presumably after a good deal of sorting had already been done. The same variety exists here, with a slight inclination to more family-related matters than in the Royal Society portion. Research is aided by a detailed hand list, where the category "Academic Material" includes some of the items printed here. Another category, "Poetry," includes a manuscript notebook containing twenty-eight poems on various subjects. Apparently Everest's effort, unapologetically recounted in her biographical portrait, "The Home-Side of a Scientific Mind," to destroy such works *in fieri* in order to steer her husband toward more serious studies related to his mathematical pursuits was only partially successful.[192]

To make the manuscripts even more accessible, UCC Library, under the direction of librarian John Fitzgerald, with the expertise of Emer Twomey and Crónán Ó Doibhlin, as well as staff members at the UCC Computer Center, has developed an online portal including high-quality digital images of each page. In addition, Anthony Durity, currently a PhD student in the digital arts and humanities program, has built an online tool called the Boole Transcriptor to encourage community engagement with the documents. I am much obliged to all these people for helpful insights that contributed to the compilation of this book. In addition, I should warmly thank the staff of the London Royal Society Library, especially Keith Moore. And while I am on the subject of acknowledgments, let me gratefully mention the name of Allen Debus, late director of the Morris Fishbein Center at the University of Chicago, who introduced me to science studies in the first place.

How the new digital Boole may affect our appreciation of Boole and the "prelude to the digital age," only time will tell.

NOTE ON THE EDITION

This collection of eight lectures gives a broad panorama of the public lecturing activity of a key figure in Victorian science who became an important protagonist of Irish intellectual life. The critical apparatus takes advantage of a notable amount of scholarship on Boole and his period that has been done since MacHale's groundbreaking 1985 book, recently reprinted. Original spellings of words, when inconsistent within the text, have been modernized, and when consistent, retained. Punctuation is untouched except for the correction of involuntary mistakes. As necessary, I indicate textual variants between copies in the London Royal Society and UCC Library.

1 ON THE GENIUS AND DISCOVERIES OF SIR ISAAC NEWTON

We are this evening assembled to receive at the hands of the noble Patron of this Institution, the bust of our illustrious countryman, Sir Isaac Newton.[1] On such an occasion it seems not inappropriate to devote a short time to the contemplation of the mind and character of that gifted individual. Through the partiality of friends, I have been requested to prepare an address on the genius and discoveries of Newton. In meeting this wish, I cannot but feel conscious of my own incompetency to do justice to so great a theme. It would not be difficult, indeed, to collect such materials as might afford a connected view of his life; or, if from popular anecdotes it were required to form an estimate of his character, the task would be easy, for with such the world has been abundantly supplied. Our present business is rather with the mind of Newton; and if, as the humblest of his disciples, I shall be thought to have succeeded in showing what were indeed the characteristics of that powerful mind, my end will be answered; or, failing in the attempt, I may hope for some measure of indulgence, when it is considered how vast a subject I have undertaken.

Sir Isaac Newton was born in the Manor House at Woolsthorpe, near Grantham, in the county of Lincoln, on Christmas Day, in the year 1642. The house in which this event took place is, I believe, now standing, and in the possession of the Turner family. In this habitation Newton continued until his 12th year, when he was sent to the free-school at Grantham. It was during his residence there, that the earliest recorded indications of talent were given.[2] His mind appears at that time to have been much occupied with mechanical contrivances, yet not so deeply as to exclude the lighter amusements of drawing, and even poetry. The scientific adjustment of the paper-kite and its appendages, and the construction of sun-dials, water-clocks, and mill-work, afforded him a philosophical past-time, which could not fail to invigorate his natural powers of invention. Even his sleeping apartment is said to have been garnished round with the untutored productions of his knife and pencil.[3]

These circumstances have been considered as indicative of his future greatness, and we might give credence to the omen, were it not that the promise of early genius has been rarely answered. In the intellectual, as in the physical constitution, a too early development is often followed by a feeble and transient maturity. The youthful occupations of Newton rather manifest the workings of an ardent and inquiring spirit, than of a mind already invested with the features of genius: above all, they exhibit a character of resolution and untiring diligence, which is in every pursuit the surest and safest road to excellence.

At the age of eighteen, Newton was sent to Cambridge, and entered as a scholar of Trinity College. Previous to this period he does not appear to have made much progress in his studies; but his mental powers now attaining, after a long repose, their natural and healthy development, soon made up for the deficiency. Without preparatory reading, he made himself master of Des Cartes' Geometry, of Wallis' Arithmetic of Infinites, and of Kepler's Optics. It has been said that on these subjects he became, while yet a student, more deeply learned than his tutors: but on what authority the assertion is made, I know not. On the resignation of Dr. Barrow, he was appointed his successor in the Lucasian chair of Mathematics, having previously obtained a Fellowship.

From this period we may date the commencement of his brilliant public career. The first subject of importance that engaged his attention was the phenomena of prismatic colours, observed a short time before by Grimaldi.[4] The results of Newton's inquiries were communicated to the Royal Society in the year 1675, and afterwards published with most important additions in 1704. The production was entitled "Optics; or, a Treatise on the Reflections, Refractions, Inflections, and Colours of Light."[5] It is certainly one of the most elaborate and original of his works, and carries on every page the traces of a powerful and comprehensive mind.

The doctrine of refraction had for some time engaged the attention of the learned, and considerable advances had been made; but it was Newton's destiny to explore a new and untrodden path, invested with a far higher interest: the Philosophy of Colours. The opinions which prevailed before on this subject were purely hypothetical, yet wanting that which is most essential to an hypothesis: the merit of plausibility. At the very outset of his experiments, he deduced that fundamental truth, that white light is not homogeneous, but produced by a mixture of other colours. As he proceeded in his researches, a series of new and more intricate truths unfolded themselves till at length his labours were crowned by the completion of that magnificent theory, which is contained in the first book of his Optics.

From the analysis of solar light, Newton proceeded to speculations of a yet higher order. The colours of natural bodies had been hitherto totally inexplicable: I do not know that a rational solution of that difficult question had ever been attempted. If attempted, it must have been abandoned in despair. A field of rich discovery was waving its untried harvest before him, and waiting for that powerful hand which should have strength to wield the sickle.

Newton began his researches by observations on the phenomena of coloured rings. These he accounted for by the theory of "Fits of Easy Reflection and Transmission"; a theory which, though it only rests on hypothesis, serves sufficiently well to explain the appearances.[6] The results of this investigation he applied by a most beautiful analogy to explain the colours of thin plates, and to determine the relation between the thickness of the plate and the colour produced. I may refer for a familiar illustration to the splendid colouring of soap bubbles: in this instance the colour is not owing to the material of which the bubble is formed, but depends on the thickness of its sides. The colour indeed is not presented until the sides have assumed a certain degree of attenuation.

On these grounds, Newton's theory of the colours of natural bodies is founded. His argument is as follows: First, The colour of a thin plate depends upon its relative degree of attenuation. Second, If a thin plate be divided into shreds and fragments, a mass of such fragments will preserve the original colour of the plate. Lastly, The parts of all natural bodies, being like so many fragments of a plate, must, on the same grounds, preserve the same colour.

The above theory is sufficient to account for some of the more vivid and beautiful of Nature's hues. The colours of feathers, insects' wings, gossamer, etc., are doubtless owing to this cause; but as a general theory of colour it is manifestly inadequate. Philosophers have been at some pains to point out its defects, though unfortunately they have been unable to find another half as general to supply its place.

Notwithstanding all the objections which may be urged, it still remains a gigantic memorial of the vastness of that mind in which it was conceived. One characteristic of Newton's mind it strongly exemplifies: the faculty of generalizing; and this, if I mistake not, was one of his chief points. In the instance before us, he commences his research by experimenting on thin plates of air: from this he deduces the germ of his theory, and to its laws he subjects, in the course of his inquiry, the whole superstructure of material things. It is true that his theory has been left imperfect; admit that in some of the applications it has failed. But at the same time we must acknowledge that in what he failed he did not fail as a common mortal, and that the marshalled intellect of Europe has vainly endeavoured to fill up the chasm!

There is in the very idea of light something so vague and intangible, that our imagination can with difficulty attribute to it an independent and material existence. Yet granting this, and assuming as our data, that under certain known circumstances, known impressions are received which we designate colour, the analysis of its primitive elements, and of the laws and effects of their combinations, would still remain a mighty problem. It is singular, that of all the subtle and mysterious agencies, light, heat, electricity, attraction, connected by one general link, and commissioned by their Author to confer upon dead matter the life and beauty of the universe, light is the only one that has yet thoroughly unfolded the harmony of its laws and submitted itself to human scrutiny. That genius which stands foremost in the triumph was the calm, patient, all-surmounting genius of Newton.

Newton's earlier optical labours were interrupted by the breaking out of the plague in Cambridge. He then retired to Woolsthorpe, and there continued two years. During that period he conceived the idea that the moon's motion might be regulated by the attraction of the earth. It is generally supposed that his attention was directed to this subject by observing the falling of an apple. If this tradition be correct, it strongly teaches us what vast effects may arise from trivial or common occurrences, when the latent energies of nature or of mind are thereby roused into action. The falling of an apple was an every-day occurrence, yet its moral consequences have been, to all human appearance, greater than the downfall of an empire. It had touched upon some hidden spring—some sleeping and folded energy: a train of thought was excited, which, though interrupted, was never abandoned, until the foundation was laid of the great science of Physical Astronomy, that science which, in its subsequent developments, has, above all others, demonstrated the economy of the universe, the capabilities of our own immortal nature, and the majesty of the Being who created them.

I have said that Newton's inquiries on gravitation met with an interruption: that interruption was occasioned by an error in his data. In his calculations he had made use of the common but erroneous admeasurement of 60 miles to a degree of the terrestrial meridian, instead of the more accurate approximations which have been since obtained.[7] The primary object of his inquiries was to compare the force of gravity in the lunar orbit with that obtaining on the earth's surface. The results at which he arrived did not, as might be expected, answer his ideas. The subject was laid aside until more accurate data enabled him to verify his hypothesis.

It is a very prevalent, but mistaken opinion, that the doctrine of universal gravitation originated exclusively with Newton. The existence of such an influence is strongly advocated by Plutarch; it is indistinctly intimated in the mystic writings of Plato. It is certain, too, that about the time of Newton's investigations, an opinion was spreading among the more profound thinkers of the age, that the celestial motions were in some measure influenced by the action of such a principle; but the prevalent and orthodox opinion was, that the planets were carried round the sun by the revolutions of an everlasting whirlpool, like straws on an eddy of water. This system was invented by Des Cartes [René Descartes], and is for that reason generally called the Cartesian Philosophy, or the System of Vortices.

In the awarding of scientific honours, it is admitted that the discovery of a new principle is due to him who first proves its existence. On this ground Newton's claim to the discovery of universal gravitation must rest. He found it an unsupported hypothesis—he left it an established truth. The discovery of this principle, with its most important applications, was announced to the world in the year 1687, by the publication of the "Principia, or Mathematical Principles of Natural Philosophy."[8] The title of the Work is such as to render it a forbidden book to the generality of readers; but since the subjects of which it treats are among the most interesting as well as the sublimest of human studies, it may not be amiss briefly to advert to its contents.

The object of the "Principia" is twofold: to demonstrate the law of planetary influence, and to apply that law to the purposes of calculation. It is true, that in certain portions of the work, investigations of a different nature are introduced, but it is possible to trace throughout them all a greater or lesser degree of connection with the main topic.

Three properties of the celestial motions had been discovered by Kepler, the astronomer, from whom they borrowed their name. By these properties the truth of a new theory was to be tested. The first of Kepler's laws (for thus they are designated), defined the orbit or path of a planet to be an ellipse; the second law determined the velocity of the body in any portion of that orbit; and the third expressed the whole time of a revolution in an orbit of known dimensions. Newton showed that these laws were the necessary results of a gravitating force according to the assumed law of its variation, and thus placed the existence of that principle beyond all legitimate doubt. The first object of the "Principia" was now attained. He had fully shown the principal motions of the planets to be such as would obtain in a system of bodies revolving by a mutual attraction varying in intensity as the inverse square of the distance. But there yet remained many important phenomena to be accounted for: the tides of the ocean, the spheroidal figure of the earth and planets, and the irregularities of the lunar orbit. All these requiring the application of his new principle, may be considered as the second object of the "Principia." To accomplish the explanation of these, some more powerful analysis than had hitherto been employed was necessary. For this purpose he unfolded the principles of the celebrated doctrine of fluxions or limits, and applied them with consummate skill in his subsequent researches. But from the imperfect form in which his new calculus then existed, the objects of its application were but partially attained. Yet his attempts, though not always, from the reason I have stated, completely successful, are never unworthy of their author. It was the struggle of a great mind with limited means, always exhibiting a mastery over them, and often rising far above them. There are instances in which he seems to have abandoned the ordinary mode of inquiry as insufficient or inelegant, and to have relied on the sole force of his genius in detecting and employing those hidden analogies which could alone surmount the difficulty.

The peculiar disadvantages under which Newton laboured will be better understood when a few remarks have been made on the nature of his proposed investigations, and the means which he possessed of pursuing them. And, perhaps, the simplest case for illustration will be found in the inequalities of the moon's motions. It is, moreover, the foundation upon which his ulterior researches on the subjects we are considering, mainly depend.

To determine with accuracy the position of the moon at any given period, had long been a desideratum in practical science. Until this object should be attained, all substantial improvements in the art of navigation, and in the accuracy of sea charts, were hopeless. Now it had been shown in the earlier portion of the "Principia," that the motions of a planetary body, when influenced by the sole attraction of another, were accurately elliptical. Such relations of velocity and time had also been assigned, as rendered its position determinable to any required exactness. Had the moon, therefore, been solely under terrestrial influence, no further search would have been necessary, and the most important branch of physical astronomy would have begun and ended with Newton. But in the actual operations of nature, such a case is far from obtaining. The moon is not only swayed by the attraction of the earth, but in some measure influenced by that of the sun. The action of the latter body has been technically called disturbance, because its effects are to disturb, or in some measure, derange the motions of the attracted body. To account for the extent and variations of every observed derangement by the theory of gravity, was Newton's first aim; to calculate the amount of the principal derangement, his second. In both attempts the profound skill of the workman and the imperfection of his means, are powerfully shown. The instrument of his research was, as I have stated, the method of limits, and the fluxionary calculus; the latter existing, as I have likewise observed, in a rude form, but susceptible of indefinite improvement. But when the same instrument was applied to the same end by his disciples in France, it had become doubly commodious and powerful, from the aggregate improvements of a century.

There was yet another disadvantage attaching to the whole of Newton's physical inquiries, which, though it gave rise to the most sublime applications of geometry, must yet be considered as having presented an insurmountable barrier to his progress, the want of an appropriate notation for expressing the conditions of a dynamical problem, and the general principles by which its solution must be obtained. By the labours of La Grange, the motions of a disturbed planet are reduced with all their complication and variety to a purely mathematical question. It then ceases to be a physical problem; the disturbed and disturbing planet are alike vanished; the ideas of time and force are at an end; the very elements of the orbit have disappeared, or only exist as arbitrary characters in a mathematical formula.

In Newton's investigations this felicitous transformation could not take place. Nature must be combated on her own grounds: the disturbing force is analyzed: its effect must be considered in every variety of position—above, below, and in coincidence with the ecliptic plane: from syzygy to quadrature, and thence again to syzygy, the same influence is to be followed, and its resulting effects determined. The everlasting wheels of the universe are before us, and their revolutions are to be traced through all the changing varieties of cause, circumstance, and effect. It is not to be denied, that this mode of investigation is attended with such complication as to render it decidedly inferior, both in power and facility, to the methods now pursued. Yet there is one respect which it possesses an advantage. Following step by step the process of Newton's demonstrations, we become more familiar with the machinery which they are intended to unravel, than if our results were immediately obtained by the discussion of an analytical formula. This consideration is, I am persuaded, of some importance to the young astronomer.

In speaking of Newton's optical discoveries, I have remarked on the power of generalizing as a prominent feature of his mind: a portion of the "Principia," on which I have now been commenting, abounds with instances of this faculty. Perhaps its most remarkable exercise may be the observed problem of equinoctial precession. In this particular case he passes from the consideration of solar influence on a revolving satellite, a point already determined, to the influence of the same power on an ellipsoidal shell encompassing a revolving planet. Of such investigations, however, sufficient has already been said.

But it was not in the power of generalizing alone that Newton differed from all other men. In this one respect, indeed, he resembles the lawgiver of an old philosophy, Aristotle. But there is this material difference, that in the latter it was the main and distinguishing feature; in Newton, it was subservient to, and perhaps resulted from, qualities of a much loftier order: vastness of comprehension, a deep and far-seeing penetration, and last and chief (the attribute of character rather than of mind), unwearied and unconquerable resolution. These were the elements of his intellectual greatness. They have stamped with immortality every page of the "Principia."

It has often been observed, that the biography of eminent literary men presents but few objects of general interest. The progress of their lives seems only measured by the order of their attainments and productions. From these we estimate the gradual rise and advance of mind and character, through every successive change, from the nursery to the grave. But in Newton's history there was one circumstance of a different order, which, from the remarkable effect it produced, cannot be passed over in silence. I allude to the well-known circumstance of the burning of his papers by the throwing down of a lighted candle. The accident was occasioned by a little dog, and called forth that remarkable expression of patience, "Oh Diamond, Diamond, thou little knowest the mischief thou hast done."[9]

The sudden loss of the labours of many years, acting on a frame already morbid with the irritation of an epidemic disorder, is believed for some time to have affected his understanding. But whatever may have been the nature of the affection, one thing is certain, the course of his life was altered. From this period he entered upon no extended field of scientific inquiry. The volume of nature was, in a great measure, relinquished, and that of revelation henceforth occupied its place. A change so important as the one alluded to, has by some been imagined to proceed from the temporary aberration which he is supposed to have experienced. Such a conclusion appears to me at best to be far-fetched. It is not difficult to conceive that illness may have deepened the tone of a naturally solemn and devotional mind. Perhaps, too, that universal satiety which mingles even with the cup of knowledge may in some measure have fallen upon him, who of any men had drank most deeply; or the advance of years, and the impossibility of restoring his lost papers, may have been of themselves sufficient to effect the change. I do not say that to any of these circumstances his theological writings are due, because all our suppositions are only founded on probability; but I would intimate that it is unphilosophical to attribute to a temporary insanity, those religious impressions which so many other causes may have tended to develop.

From the period of this critical illness until that which terminated in death, the current of his days glided calmly and evenly along. All disputes on the merit and originality of his discoveries were now at an end, and he was daily receiving those tributes of honour and affection which had long been acknowledged his due. In the year 1699 he was appointed Master of the Mint; subsequently the honour of knighthood and the Presidency of the Royal Society were conferred upon him. The intervals of his public duties were chiefly employed in theological enquiries, yet so as not to neglect the subsidiary studies of chronology and history. All these sciences he has enriched. His chronology, and the occasional speculations in mathematics which he has left us, are sufficient to show that the vigour of his early years had not departed. His observations on the book of Daniel and the Revelation of St. John will remain as monuments of his vast learning and deep sagacity so long as the pages of the prophet and the evangelist shall continue to retain their claim on the interest and veneration of mankind.

We cannot but consider this later portion of Newton's life as highly and singularly happy: all that can make old age honourable he possessed, with scarce a shadow of its dotage and infirmities. He was not a father, but the natural affections were in him expanded into the broad principle of universal philanthropy. Subdued passions, moderate wealth, and the much-loved blessing of peace, all tended to smooth and to illumine the rugged path of declining life. If desire of fame had been the meteor of his youth, it could not now disturb his repose, for he had long been at the summit of all earthly ambition. If the recollection of the "single talent well employed" be attended with a pleasure, surely that pleasure must have been felt in its keenest relish by him, who had received from his Maker ten golden talents, and well employed them all. These are the materials of happiness, and all these were possessed by Newton. But more than these, the support and solace of his faith, the prospect of future happiness which grows brighter as all other prospects decay, these were unalienably his. And though genius has too often been a wandering star, the minister of licentiousness, or the associate of scepticism, in his life we have ample testimony that such is not a natural or a necessary alliance.

Nor perhaps is there less to admire in the high excellence and unblemished purity of his moral character, than in that halo of philosophical glory which has gathered around his name.

In the year 1722 Newton felt the first inroads of that fatal malady, which at length terminated in death. By rigid economy of diet he was enabled to ward off for some years the final catastrophe, and to enjoy long intervals of health. His disease was of a trying and painful nature, and encouraged no hopes of an ultimate recovery. In the February of 1727 it assumed a more virulent form, and indicated, with no dubious symptoms, the rapid approach of death. The paroxysms of the disorder were now frequent and violent, and during their continuance the sweat of agony often started upon his brow. He did not complain. Newton never murmured at the severity and length of his affliction. His intellect continued bright and settled until within a short period of his dissolution. On the evening of Saturday the 18th of the following March, he fell into a state of insensibility; on Monday the 20th, Sir Isaac Newton was no more.

He lies buried in Westminster Abbey: a monument is there erected to his memory, bearing a Latin inscription, of which the following is a translation, as literal as the transfer of idiom will allow.

Here lies SIR ISAAC NEWTON, KNIGHT,
Who, with a vigour of mind almost superhuman,
Guided by the light of his own Mathesis,
First demonstrated the motions and figures of the planets,
The paths of the comets, and the tides of the ocean.
He thoroughly investigated
What none before had even suspected,
The different varieties of the rays of light,
And the properties of the colours thence proceeding:
A diligent, sagacious, and faithful interpreter
Of Nature, of Antiquity and of the Holy Scriptures,
He asserted in his Philosophy the Majesty of God,
And exhibited in his conduct the simplicity of the Gospel.
Let mortals rejoice
That there has existed such and so great an ornament of the human race.
Born 25th Dec. 1648, died 20th March 1727.[10]

We cannot but feel struck, on reading this inscription, with the variety of talent, and depth, and universality of attainments, which it is intended to commemorate. Contemplating even now the efforts of his transcendent genius, we can scarcely tell what field of knowledge, human or divine, he has most assiduously cultivated. Independent of those greater productions of which a sketch has been attempted, his career was marked by a succession of brilliant discoveries, inferior in importance to these, yet many of them singly sufficient to immortalize his name; in some, descending from the higher walks of science, to illustrate its earliest principles; in others, he has soared beyond the spirit of his time in striking out new paths of inquiry, and anticipating the discoveries of a future age.

But the arena of his boldest triumphs was the science of Astronomy. In the very choice which thus directed his inquiries there was something most auspicious for his fame. Those bright and distant worlds, whose laws it was reserved for him to investigate, have ever been the objects of human curiosity. In their silent and eternal courses they have received the idolatry of a hundred generations. Through the annals of human superstition their influence has been ever predominant, presiding by some mysterious and fearful agency over the chances and calamities of life. That feeling of dread with which they were once regarded is past; a better philosophy has dispelled the terror, but it has not diminished the interest; it has taught us to consider them not as omnipotent over the fortunes and interest of earth, but as individually possessing an interest of their own, the abodes of other forms of organic life, of other orders of intelligent existence. Such are the prospects of modern astronomers: they are bold, yet scarcely conjectural. In the very circumstance of their being admissible, we recognise the sublimity of the science. There is a high and melancholy pleasure in reading, even on a monumental stone, the records of its great establisher, NEWTON.

The very pertinacity with which error retains its hold, is one of the strongest arguments for the final and eternal establishment of truth: it results from a natural fear, that in the wreck of received opinions, the very foundation of credibility should be destroyed, and mankind a second time involved in darkness and uncertainty. For this reason, perhaps, there have been few of ancient kingdoms, which have not bequeathed to other times, a faith, or a philosophy, more enduring than themselves. The Coliseum and the Acropolis are in ruins, but the philosophies which sprung up beneath their shadows are yet deep in the tide of human opinions, still influencing, with an unseen but mighty influence, the character of the age. The dark creed of the ancient Persian is yet descending from sire to son in the sacred annals of the Guebres; and the faith and fame of Zoroaster are yet triumphant against the desolation of his country, and the sword and the Koran of its Mahometan oppressors. But more especially is this truth to be observed in the records and remains of more ancient dynasties—in patriarchal Assyria—in sepulchral Egypt. The long succession of their kings and warriors is now doubtful or forgotten—the colossal relics of their primeval architecture are daily mouldering; but their sombre religion—their wild astrology, originating in the days of their greatness, are perpetuated when the very shadow of that greatness is no more. If, from the history of false and discarded systems, we may pass by analogy to the more enduring influence of truth, we shall perceive how high above the chances of time and vicissitude, the pedestal of Newton's immortality is founded.

Among different nations, and in the succession of ages, the posthumous honours of genius have proportionally differed. There has been a period when the author of a "Principia" would have been deemed more than a mortal: that period of idolatry is past, and with it have passed the honours of its earth-born Divinities. We do not in this day consecrate the shrine or the temple in adoration of the dead; but the tribute which we now pay to the memory of Newton, shall be the tribute of all ages, and of climes—the admiration of his talents, and the imitation of his virtues!

It is now time that I should conclude. Perhaps I have erred, in reverting at such length on an evening of festivity to a theme of more solemn interest, the recollections of Newton. If the veneration which I bear for that great name has thus far misled me, I know I shall be forgiven; but I am encouraged to hope that the hour we have spent has not been wholly barren of instruction. The details of a sage's history, no less than the productions of his retirement, are pregnant with golden wisdom; but it is that homely and practical wisdom, which is accessible to all its suitors. When this marble shall call up to our memories the career of Newton—his patient struggles, his eternal triumphs—it shall not be without a responsive chord from our own bosoms. If, in that silent admonition, one human spirit shall be awakened to its great duties, to suffer with fortitude, or triumph with humility, to expand with science, or warm with philanthropy, that marble shall not have left its native quarry in vain; but I am persuaded that this sacred influence will not be confined to an individual breast, but breathe into each of our bosoms; so shall we show more effectually than words can express our estimation of the value of the gift, and our gratitude to the noble giver; and I cannot but regard it as a most auspicious omen, when I see on this interesting occasion the distinguished representative of an illustrious house, cooperating heart and soul in that noblest work of British patriotism, the education, the enlightenment, and the happiness of his fellow countrymen.

2   ON THE CHARACTER AND ORIGIN OF THE
ANCIENT MYTHOLOGIES

There are few subjects on which the opinions of learned men so widely differ as on the question of the origin of the ancient mythologies.[1] This is to be attributed partly to the nature of the inquiry itself, which does not proceed on the ordinary ground of historical evidence, but involves all those difficulties and perplexities which beset our path when we would advance beyond the limits which separate the ages of history from those of fable; and partly too, perhaps, to the influence which preconceived notions and peculiar habits of mind have exercised upon their judgment. The question is difficult not only from the paucity of the evidence but also from the contradictory character of that which we do possess. Very few of the ancient writers agree in the representations on this subject, even though living in the country whose opinions they profess to relate. And when we compare the statements which have been transmitted by Greek and Roman writers on the religious opinions of nations with which they were less immediately connected, we meet on every hand with discrepancies so great that the attempt to reconcile them is a hopeless task, those who have written much upon the subject generally having started with some theory of their own to which they have endeavored to bend the facts they met with. The attempt to reduce to some single principle the various and discordant relations of mythology is however little likely to meet with any success if we may judge from the past. Such attempts are however not without their use. From a comparison of the results, conducted with as little partiality as possible, we shall be more likely to obtain just ideas, and so arrive at the truth, than by following any other course.

In tracing back the history of the religious institutions of pagan antiquity, we sooner or later arrive in each instance at a period in which the

rites and instruments with all the attendances of religion assume a degree
of simplicity unknown to later ages. On the rude altar of turf or unknown
stone, the idolatrous sacrifices of the early world were offered. In caverns
and on mountain peaks, the oracular worshipers of its divinities were
heard. In rude enclosures of direct stones, and under the open canopy of
heaven, the ceremonies of its worship were performed. Superstition had
not yet called in the aid of architecture, and sculpture was altogether
unknown. So long as the primeval forms of elemental worship continued, their necessity was unfelt. The sun, the host of heaven, and the great
phenomena of nature were regarded either as objects of adoration themselves or as the visible manifestations of unseen divinities. Perhaps too it
might be generally thought that a greater sacredness attached itself to
those objects which had not been moulded by human skill, and that a
shapeless block of granite, washed from its original seat by mountain tor-
rents, was a fitter object of veneration than the more regular forms which
the chisel might have produced. That a sentiment of this nature did exist
among nations advanced to a very high degree of civilization we know,
and it is not improbable that in earlier ages its influence may have been
more widely acknowledged, and may have given rise to that taste for
rude and massive architecture which we continually discover in the
remains of the Cuthite and Pelasgic tribes.[2] Perhaps the simplest and
most elementary form of mythological worship is that which prevailed
in Persia, and which united the adoration of fire, to the doctrine of two
eternal, opposite principles. This, with some unimportant additions, is
the scheme taught by Zoroaster and the Magiian priests. Of the nature
of the Persian worship before the era of Zoroaster, we have no very definite information, but it is probable that so far as concerns the characteristic features of the Persian mythology in the adoration of the sun, but
little change was introduced. Indeed all the circumstances attending the
celebration of those rites bear the evidences of extreme antiquity. It was
usual to offer the sacrifices on mountaintops, or in those rude and primitive temples, by what the Greeks termed ὕπαιθρα, which were constructed without roofs, so that the ceremony might be performed in the
open air. No inconsiderable portion of the Persian ritual consisted in
nocturnal vigils performed by the Magi, in caverns either formed by
nature, or hollowed out of sides of the rocks of mountain precipices with
which that country abounds. In those gloomy recesses, they performed

the more secret rites of their religion. The frantic howls of exclamations with which they were attended, and the mournful character of hymns which were there chanted, gave a solemn terror to the localities and inspired with deeper and more superstitious reverence the devotees of that fearful worship.[3]

The hypothesis of two eternal and intelligent principles standing in direct antagonism to each other and producing, by their mutual conflict, the mingled good and evil which we see around us is evidently the hinge on which the whole system of Zoroaster revolves. To these imaginary beings, the names of Oromiede and Ahriamanes were respectively given, and between [these], a third (Mythias) was supposed to exist, to whom a kind of mediatorial office was assigned.[4] Though represented under a human shape, he was supposed to be identical with the sun, or deity of fire, and as such received a principal share of divine honors. The worship of this fabulous being was attended with circumstances of great cruelty. No one was admitted to the priesthood until he had undergone an ordeal calculated to deaden and destroy all the gentle feelings of his nature, and to steel his bosom against kindly influences of pity or affection. This terrible process of initiation consisted in a gradation of torments [over several] days to the number of ten, in the course of which, as we are informed, many expired, and others came out from the trial crazed and shaken in their intellects, blunted in their moral susceptibilities and prepared to inflict on other victims the sufferings they had endured themselves.

Of the first of these forms, i.e., in the adoration of Eternal Nature, we have a remarkable example—the Sun Worship of the early Persians. The account given by Herodotus, who flourished in the fifth century before the Christian Era, is to the following effect. "They have neither statues, temples nor altars; the use of which they censure as impious, and a gross violation of reason. Their custom is to offer from the summits of the highest mountains, sacrifices to the expanse of heaven. They also adore the sun, the moon, earth, fire, water and the winds, which may be termed their original deities. To these they sacrifice without altars or fire libations or instrumental music, garlands or consecrated cakes. The head of the supplicant is surrounded by a tiara, adorned with myrtle, and when the victim has been divided, the Magiian priest, always in attendance on such occasions, chants the primeval origin of the gods."[5] Subsequent writers describe the sun and the principle of fire as the more especial objects of the religious regards of this nation and mention the construction of ὕπαιθρα or columnar temples, of a circular form and open to the sky, in which the observances of their ritual were performed, the centre being occupied by the sacred fire. That they moreover called in the aid of sculpture and employed images and symbolical representations is evident from existing monuments. Particular forms of the religious system of the ancient Persians are found scattered through various and widely distant nations. The adoration of the sacred fire was carried by the Etruscans, who were a Lydian colony, into Italy, and introduced by Numa into Rome. Hence the institution of the Vestal Virgins whose office it was to tend the flame. The analogy between the roofless temples to which allusion has been made and the columnar structures of the Druids will at once be recognized.[6]

But the most striking feature of the Persian theology, when more maturely expounded, was the deification of the two antagonistic principles of Good and Evil. Ormusa, we are told, the author of Good, created an egg, a symbol of the universe, in which he inclosed certain genii or spirits, similar to himself, and typical of the moral virtues, but Ahriman, born of darkness, broke the shell of this imaginary world and introduced an equal number of genii, partaking of his own character.[7] Hence arose a contest for sovereignty, in which the triumphs of the benevolent deity are followed by the spread of virtue and happiness in the world, those of the malignant one by disorder in human society, and convulsions in the frame of nature. Finally, however, the principle of Good is to assert its superiority. The bodies of mankind shall then be freed from pain and yield no shadow; and evil and darkness disappear forever. It seems to be a matter of dispute among the learned, whether this doctrine formed any part of the original belief of the Persian Magi, or whether it was the product of later speculation, perhaps three or four centuries before the Christian era. From the circumstance of its not being mentioned by Herodotus, we may conclude that it had not in his time assumed a mythological form, even if as an abstract opinion it was then in existence. If this view be correct, the religious notions of the Persians underwent a regular process of development, in the first stage exhibiting the characteristics of a very pure Druidism, the simple worship of External Nature; in its second acquiring images and symbols, with probably some fragments of mythological history; in its third assuming the form of a philosophical allegory, investing with personality those abstract ideas which had arisen in the human mind, from an attempt to explain the seeming contradiction in the moral government of the world.

I have dwelt at some length on the religious doctrines of the ancient Persians, particularly as regards their mythological theory of Good and Evil, because I think it affords a clue to the interpretation of various passages in the mythologies of other nations. To the elucidation of this point I shall devote some observations in a subsequent portion of this essay.

The religious system of ancient Egypt, to the consideration of which we shall now proceed, appears to have combined the grosser elements of Hero worship with a speculative doctrine, bearing great resemblance to that we have just dismissed. It was to an eminent degree a symbolic religion and affords a startling illustration of the abuses to which an indiscriminate use of symbols may lead. Whatever may have been the secret philosophy of the priests, to the apprehensions of the vulgar they offered but a rabble of bestial divinities: the ichneumeon, the sacred beetle, the cat, the crocodile, the bull Apis.[8]

> A crew, that under names of old renown,[9]
> Osiris, Isis, Orus and their train,
> With monstrous shapes, and sorceries abused
> Fanatic Egypt and her priests, to seek
> Their wandering gods, disguised in brutish forms,[10]
> Likening their Maker to the grazed ox.

If Diodorus Siculus is to be credited, the original deities of Egypt were the sun and moon, under the names of Osiris and Isis, with certain primordial powers of nature, nearly identical with those to which, according to Herodotus, the worship of the early Persians was directed.[11] And this view appears to derive some confirmation from a subsequent remark of the former author, that it was a dispute among the priests, whether the first of the dynasty of gods, which in the beginning ruled over Egypt was Helius the Sun or Hephaistos the god of Fire. I mention these observations of the historians here, because they seem to indicate the existence of a period in which the religion of Egypt had not assumed that mythological and highly complex form which in after ages it exhibited; had not in fact extended beyond the adoration of External Nature. This stage there is reason to think has been common to all nations, more particularly as respects the worship of the sun. Man, losing the knowledge of a Creator, sought first to supply his place by the most resplendent object of the material creation.

Of the Egyptian mythology, as handed down by the priests of later ages, we have the fullest account in Plutarch's treatise on Isis and Osiris. From this, compared with such portions of the writings of Diodorus and Herodotus, as are devoted to the same subject, I have compiled the following abstract, omitting however such minute details as do not assist in its interpretation.

We are then told, that while the world was yet immersed in ignorance and barbarism, Osiris and Isis, sprung from a line of gods, swayed the sceptre of Egypt. Imprinted with pity for the miserable condition of the human race, they formed the benevolent design of imparting to them the knowledge of those arts by which life is adorned and elevated them from a condition of rapine and hostility to the enjoyment of peace and social order. Beginning with their own subjects, they persuaded them to relinquish the practice of cannibalism to which they had been before addicted, instructed them in agriculture and taught them the use of various fruits and herbs, which until then had been suffered to grow wild from ignorance of their value. Of these, Isis is said to have introduced the cultivation of grain, and Osiris that of the vine. Subsequently they established laws, built temples, instituted religious worship, and a priestly caste, to whose maintenance a third of the lands was appropriated. Egypt thus reclaimed from barbarism, Osiris determined to extend the same benefits to the remainder of the world. For this purpose, leaving Isis in conjunction with Hermes, to hold the scepter in his absence, he proceeded at the head of an army on his philanthropic mission. This army which accompanied Osiris was not like the conquering hosts of Ramases [i.e., Ramses] or Semiramus, but consisted of reapers and vine dressers, of musician and poets; and the arms which they carried with them were the sickle, the pruning hook and the lure. In this expedition he is said first to have visited Ethiopia, thence he pursued his course through Arabia and India and finally entered Europe, dispensing the seeds of civilization to the rude tribes by which that quarter of the world was then inhabited. Wherever he went he taught the culture of grain, and if soil and climate permitted, that of the vine. If, says Diodorus, he arrived at any region too inhospitable for the latter purpose, he instructed the inhabitants how to prepare from barley a liquor not much inferior to wine, in strength and flavour.[12] Finally returning to Egypt, laden with the spontaneous offerings of nations whom he had subjugated by kindness and persuasion, he received from his grateful subjects the tribute of divine honours.

Before we proceed to consider the somewhat tragical sequel of this pleasant fiction, it may be appropriate to observe that similar passages occur in the mythologies of various other nations of the ancient world. The institution of laws and religious rites, and the discovery of grain, and planting of the vine, are events which the patriotism or national vanity of almost every people has attributed to some deified individual of their own race, and the opinion is usually conveyed in some narrative not unlike that of the fabulous expedition of Osiris. Of such examples it will suffice to notice that of Bacchus and Ceres, in the Greek mythology, whose history seems almost a repetition of that of the Egyptian deities. Bacchus with an army of his votaries travels through Asia, introducing the knowledge of the vine, and never exerting his power in an offensive form, except to those who derided his inventions or refused to acknowledge his divinity. This parallel, as well as that between Ceres and Isis, has been drawn out at great length by Diodorus and is so close as to have induced many to suppose that the worship of the two Greek divinities was originally brought from Egypt.[13] Whether such was or was not the case, is not very easy to determine. All that we are at present concerned to notice is the general fact of the existence of such analogies.

We now approach that portion of the mythological history of Osiris and Isis which has most tasked the ingenuity of the learned. After the return of the former to Egypt we are told that his brother Typhon, envious of the reputation which he had achieved, determined on his destruction. This wicked design he effected at a banquet, and inclosing the body of his victim in a chest or ark, committed it to the Nile, by whose stream it was carried down to the Tanaitic mouth, and there entangled among the aquatic plants which grow along the margin of the river.

After many fruitless attempts to discover the body, Isis at length succeeded in obtaining possession of it, but was unable to secrete it from Typhon, who again meeting with it, divided it into fragments and distributed them over Egypt. To recover these, the goddess constructed an ark of bulrushes, in which she sailed over the fens, and having collected the greater part of the mutilated limbs of her husband, she, according to one account, buried them separately in the different names of Egypt, according to another, interred them together in an island of the Nile near Memphis. After this event we are informed that Osiris rose from the dead and instructed his son Orus to avenge his death. I shall not pursue this absurd fable further, nor indeed should I have dwelt upon it thus far, but that some acquaintance with it is absolutely necessary to the exposition of these views, which learned men have been led to form of the origin and interpretation of the religious system of the ancient Egyptians. What these views are, we will now proceed to consider.

The learned Jacob Bryant, and I believe the majority of classical antiquaries, supposes the god Osiris to have been no other than the patriarch Noah, and his mythological history a corrupted form of the tradition of the Deluge.[14] Into the arguments by which this view is supported, it would be impossible here to enter at length, and I shall therefore briefly allude to them. The first is drawn from an examination of that history, the death and resurrection of Osiris, and his first and second life on earth, which are compared to those portions of the life of Noah, preceding and following the Deluge; finally the planting of the vine and the establishment of religious worship, events which in the Scripture narrative are both attributed to the patriarch, who after the subsiding of the waters "builded an altar unto the Lord" and "began to be an husbandman and planted a vineyard."[15]

The second argument is founded on etymology.

The third argument is drawn from the symbols which are found among the ruined temples of Egypt or are known to have been employed in the religious mysteries. Of these we may especially notice the lotus flower, a species of water lily, whose expanding cup floats on the surface of the Nile, and rises with the flood, and is regarded as the emblem of preservation from that element. This meaning it undoubtedly bears in the Hindoo mythology, which in many points exhibits a close resemblance to the Egyptian. The Hindoo god Satyavatra [i.e., Satyavrata], whose history is undoubtedly taken from the Scripture account of the Deluge, or is itself a corrupted tradition of that event, is commonly represented as floating on an expanded ocean in the cup of the lotus, and this symbol was also common in Egypt in connexion with the god Osiris. Other symbols have been added, particularly the coldcasia [i.e., colocasia] or sacred bean, the seed of which resemble[s] an ark, carried in the Isiac and Osirian processions. It is not however my intention to pursue this train of illustration further, especially as much license has been given to imagination by some who have engaged in the inquiry. Abstracting however for the influence of fancy, the weight of evidence will I think still preponderate in favor of those who maintain, if not the absolute identity of the Egyptian deities with the patriarch of the Scripture history, at least the opinion that the symbols with which they are associated bear some such reference.

But a difficulty is here met with, which does not seem to have been sufficiently weighed by the framers of this hypothesis. It is that in the explanations which have been given by classical authors of the Egyptian mythology, no allusion whatever has been made to that theory, which resolves so large a portion of its history and symbols into the memorials of a deluge. From the treatise of Plutarch on Isis and Osiris to which I have before alluded, we learn what were the opinions entertained in his day, as well by the Egyptian priests as by Greek inquirers on this subject. Some of the most important of these I shall briefly enumerate.

The first explanation represents Osiris as a personification of the sun, Isis of the moon, [and] Typhon of darkness.

Another explanation makes Osiris the personified Nile, the parent of Egyptian fertility, Isis, the soil rendered productive by the fertilizing slime of the river, [and] Typhon, the sea in which its waters are lost. We may add, that to the Egyptians, until a comparatively late period of their history, the sea was an object of especial dread and aversion.

The last explanation I shall notice, regards Osiris and Isis as the principles of moisture and fertility, [and] Typhon that of drought and barrenness.

Now on comparing these interpretations together, and abstracting those qualities in which they agree, we are led to the conclusion that Osiris and Isis represent generally the deified idea of physical good, as exhibited in surrounding nature, and Typhon, the personified idea of physical evil. Similar to this is the conclusion to which Plutarch himself arrives, who maintains that the Egyptian mythology, like that of the Persian, shadowed forth a theory of the origin of good and evil, under the form of a contest of antagonistic principles. But we may here ask, does not this view of the subject militate with that which represents the Egyptian mythology as founded in traditions of the Deluge? To this question I think a satisfactory answer may be obtained from the following considerations.

Every system of religious belief which has usurped a permanent influence over any section of mankind has attempted to offer some solution of the question, whence arises good, and whence evil? This problem, the most difficult within the range of human speculation, is yet the first which the speculative mind attempts to solve. There is a remarkable difference between the ways in which man seeks to investigate natural and moral truth. In the study of the former he begins at the right end. He does not make it his first essay in Optics to construct an achromatic lens, or in Astronomy, to elucidate the perturbations of a planet's orbit. But in metaphysical inquiries, it is always the more difficult question which comes first. Thus the origin of evil was one of the first problems with which the human mind was engaged. Baffled in all its attempts to comprehend a subject with which it is not less morally than intellectually unfitted to grapple, it gives a license to imagination, and seeking to explain by fiction what it cannot account for by reason, constructs to itself a mythology. The materials which it employs either have their source in its own emotions or conceptions, invented with an outward personality, or are derived from traditions of real events. The latter I conceive to have been the way in which diluvian records and symbols were introduced into the religious system of the Egyptian priests. Nothing could indeed be more natural than that in constructing a mythological theory of the origin of evil, they should borrow some of their illustrations from an event, which, from the widespread devastation which it occasioned, might come to be regarded as a temporary triumph of the principle of evil. In the formation of the Hindoo mythology, we know that this view prevailed.

It is thus seen that the Egyptian mythology like that of the Persians underwent a regular and systematic development. In its first stage, it acknowledged no other divinity than the sun and moon, the host of heaven, and the Nile, that mysterious river, which deriving the supply of its waters from the rain falling on mountains, distant not less than 2,000 miles, distributes abundance over a land which knows no showers, and which but for it would be an arid desert. On this foundation we also see that the subsequent popular mythology of the country, a species of Hero worship, was erected, the original objects not being laid aside, but the ancient worship of the stars blended with that of earthly divinities. I do not know where to borrow a finer illustration of this change than from the language of a Greek writer who, quoting the opinions of the priests on their deified Heroes, observes "their bodies lie interred with us, but they have inscribed on their fame the title of Divinity, and their souls have become stars shining in heaven."[16] Here we have a distinct statement of the way in which the two forms of polytheism are united. The star is made the symbol of the god, and the worship which it before claimed to itself is divided with another. At the same time they are made to blend with a speculative theory, and both merge into personified ideas.

It will be here in place to offer some remarks on the Egyptian ritual and religious establishment. I have already observed that a third of the lands was allotted to the priests, who besides the duties necessarily devolving on their order, filled the most important offices of the state.[17] From them were selected the magistrate judges, and they had the imposition and the levying of the public taxes. Their customary dress was a robe of white linen. On the head, which was shaven, they wore a high crowned bonnet, not infrequently wreathed with the figures of asps. In their sacerdotal character they presided over sacrifices, practiced divination, and composed the funeral records and eulogies which are found inscribed in the monuments. And that the last duty was regarded as of great importance may be shown by the most casual inspection of the Egyptian rooms in the British Museum. The sarcophagi will be there found to be covered within and without with hieroglyphic paintings. The sacrifices were of various kinds. Originally there is reason to suppose that they consisted merely of the fruits of the earth, but after ages the immolation of animals, and even of human victims, was practiced. On the forehead of the victim a seal was set, a solemn curse was then pronounced upon the head, after which it was cut off and thrown into the Nile, or sold to strangers. To the god Typhon, as the principle of evil, where annually offered appeasing or hurting sacrifices, the victim in this case being usually an ass or a swine. The latter of these animals was regarded by the Egyptians with a degree of abhorrence second only to that in which it was held by the Jews. Between the two nations there was however the difference, that while the Jews abstained entirely from its flesh, among the Egyptians it was prescribed to be eaten as a religious duty, on the location of those yearly sacrifices to which I have just alluded. Among the sacred animals, the cow, symbolical of Isis, and the bull, of Osiris, were held in the highest veneration. But to such an extent was the practice of animal worship carried, that there was scarcely a species of quadruped or reptile on the face of the country which did not in one place or another receive divine honors. Modern research is fully confirm[ing of] the otherwise incredible statements of ancient writers on this subject. In the heart of the pyramids have been found the preserved relics of the once living idols of this besotted people. In the ruins of Memphis are some long narrow cavities believed to have been made for the reception of the embalmed bodies of the sacred crocodiles. A strong tendency to superstition still characterizes the Egyptian mind. I remember to have read of some African traveler who seeing a peasant on the bank of the Nile kneeling before a picture, had the curiosity to approach, and found the object of his worship to be a rude figure of Napoleon Bonaparte whose military career in that region had left the impression that he was a god or demon, whose wrath was to be averted by prayer.

In the course of the preceding observations I have once or twice had occasion to refer to the opinions and traditions yet prevalent in Hindustan. The form of paganism developed in that country at the same time that it is one of the most ancient in the world, is almost the only one which has descended without material change to the present. Over a large portion of the south of Asia, it still rules with a scarcely diminished sway. I shall therefore claim your further indulgence while I offer a few brief and general observations on the subject. To enter into minute details will be neither profitable nor interesting.

The Hindoo theology, while it teaches the existence of an almost infinite number of gods, genii, and demons, recognizes only the supremacy of three: of Brahma as creator, of Vishnu as preserver, of Shiva as destroyer: or as some are more willing to suppose, of one original deity, the neuter Brahma under three different manifestations. The functions of Vishnu and Shiva do not always appear to have been of a determinate kind, and their titles are not to be supposed to indicate the invariable character of their operations. To Brahma the work of creation is assigned, on which he is supposed to have been engaged [for] several millions of years. The method he employed is almost too absurd for description; and it will suffice to notice that the four great castles, or hereditary divisions of the Hindoo people, those of the priests, the soldiers, the husbandmen and the servile laborers, are supposed to have respectively sprung from his head, arms, thighs and feet. Some time after the completion of his task, Brahma fell asleep, and a demon took the advantage of his slumbers, to immerse the world in a deluge. On this occasion, the god Vishnu appeared in the form of a fish, and bringing an ark through the waters, preserved the king Satyavoltra [i.e., Satyavrata] and his family, by whom the earth was repeopled. This incarnation of the god in the form of a fish, and in the character of Preserver, is called in the Indian Scriptures his first avatar. Nine such avatars of Vishnu are recorded to have taken place, in most of which the bodily shape was assumed, for the purpose of averting some great calamity about to befall the human race, or of punishing impiety or tyranny. In the eighth avatar he became Krishna, a being whose history shows him to have been no other than the personification of the grossest of the animal passions, and who in that character is still a favorite object of adoration in India. In the ninth he became Boodh, whence the origin of that form of religion called Boodhism, which, though denounced as heretical by the Brahmins, has overspread Ceylon and the further Indian peninsula, and it said to constitute the belief of a large portion of the inhabitants of Tibet and China. It is I believe the opinion of many orientalists, that Buddhism at one period prevailed in Hindoostan, over the present form of Brahminical superstition. It is even mentioned by Clement of Alexandria, who flourished at the close of the second century of the Christian era. Ultimately it was expelled by force of arms. It is a common opinion that the god Boodh was a religious reformer, whose doctrines, not being sufficiently favorable to the domination of the priestly case, arouse the fears of the Brahmins, and led to the expulsion of his followers from Hindustan. The 10th avatar of Vishnu, which is yet to be revealed, has been made the burden of Hindu prophecy. He is to appear as a conqueror, and vindicate his sovereignty through the world.

It seems to have been in a particular degree the fate of India to suffer from religious wars. The expulsion of the Boodhist I have already mentioned, and the details of the Mohammedan are well known to readers of history. Between the two great sects, the worshippers of Vishnu and those of Siva, a deadly hostility subsists, even to the present day, and but for the dread of British power, armed hordes of fanatics might even now be carrying fire and desolation through the land.

Of the actual development of the Hindoo mythology there can be no
doubt, that it was similar in its progress to that which we have traced in
the coordinate systems of Persia and Egypt. The most ancient of the
sacred books of the Hindoo agree in representing the sun and the elements as the primeval deities. To the luminary of day, every Brahmin
offers his morning devotions. Both as a deity and as the symbol of deity,
that orb is still reverenced through India, and there are some districts in
which no other worship has been introduced. Of Hero worship it would
be idle to multiply examples. The worship of personified ideas is illustrated as well in the supreme triad of deities, in the characters of Creator,
Preserver and Destroyer, as in a host of minor divinities whose functions
cannot here be particularized. What is the predominant idea of the Hindoo theology, it is not easy from so confused a mass of allegory and fable
to determine, but everyone will recognize the resemblance between the
Vishnu and Siva in their characters of Preserver and Destroyer, and the
Ormusa and Ahriman, the good and evil principles in the Persian creed,
or the Osiris and Typhon of the Egyptian. Thus the analogy proposed is
complete in all its parts. That it may be extended to the mythologies of
various other nations might be easily shown, but a few general observations must suffice. It is so universally admitted that the primitive form of
idolatry in all those nations of the ancient world with whose history we
are sufficiently acquainted, included the worship of the sun as its most
important element, that I deem any further illustration of this point as
almost superfluous. I shall only observe that it has been found in the New
World as well as in the old, and that the kingdoms of Mexico and Peru
when visited by the Spaniards, were the seats of a splendid but despotic
and bloody superstition, of which sun worship was the prevailing feature.[18] Of Hero worship, the traces are fully coextensive. In the Greek and
Roman mythologies it is particularly conspicuous. The Jupiter and Her-
cules, the Apollo and Bacchus, the Diana and Proserpine, even in the
language of the priests who appeased them by sacrifice and the poets
who flattered them in song, are no other then deified mortals. Nor less
universal was that worship of personified ideas, which in its most important form, we have traced through the mythologies of India, Persia and
Egypt. The origin and operation of evil, both in the moral and physical
world, we have there found represented by a contest of antagonist powers.
Now something of this kind is met with in almost every mythology with

which we are acquainted. In the Greek and Roman, we have the Titanic war, and the contest of Python and Apollo. The Scandinavian consists of little more than a series of battles, in which Odin, Thor and their fellow gods are engaged with the races of giants and evil spirits. It is quite possible that some of these fabulous contests may have been derived from the source of tradition, but this does in no way vitiate the argument. To the popular mind, they may still have outshadowed a rude theory of the moral government of the world, and supplied what might otherwise have been wanting in the claims of the favorite divinities to the supremacy of the skies.

In connection with this topic, some notice may here be taken of the celebrated ophite, or solar serpent heirogram, which has been made so fruitful a theme of speculation. This figure consists usually of a circle with one or two serpents variously combined. Sometimes the circle is represented as winged. More than one explanation of this remarkable symbol has been given, but the simplest appears to be that it represents the fundamental notion of the later Persian theology, the contest of two opposite powers for the sovereignty of the universe, or a supreme being under a twofold character. The first example I have to offer is from China, where it is said to be very commonly employed as a religious emblem. There can be no doubt as to the interpretation of this figure. The second example is copied from the ruins of an ancient city in Persia. Similar representations are found on the porticos of Egyptian temples and among the ruins of Peru and Mexico. They have even been found tattooed on the bodies of savages. Among the Greeks, the caduceus or wand of Mercury, and the Medusa's head, are believed to have been particular examples of this figure. To this it may be added, that it is a disputed point among antiquaries, whether the construction of the Druid temples was not guided by a reference to the same design. Such was the opinion of De Stukeley, and it has since been advocated by Jacob Bryant, by Sir R. C. Hoare, author of a descriptive account of ancient Wiltshire, and by Mr. Deane, in his learned and ingenious treatise on serpent worship.[19] Popular tradition favors this view, but on the other side it may be observed that in perhaps the majority of instances, the resemblance is very slight, or does not exist. The only case with which I am acquainted in which the analogy is clear, is that of the temple of Avebury in Wiltshire, which consists of a circular and closure, with a long winding avenue of parallel ranges of stone. In Stonehenge, the circular enclosure is found without the avenue. In the enormous temple of Carnac in France, there remain the parallel ranges of stone, winding over hill and dale to a length of nearly 8 miles, but of the enclosure, no traces whatever are discovered. Lastly, in various examples of Druid structures, the avenues are straight.

I am not sufficiently acquainted with British antiquities to be able to offer an opinion on the merits of this controversy, [but] possibly someone present may be disposed to offer his views upon it. Before dismissing the subject, I will just take the opportunity of observing that history is by no means destitute of analogies to show that the constituent parts of a sacred edifice have in various nations been regarded as symbolical. Among the builders of our ancient minsters such notions were prevalent and almost every portion of those venerable piles was invested with a figurative significance. These opinions of our Catholic ancestors are illustrated in a rich vein of poetry in Moiles' State Trials.[20]

It is a remarkable circumstance that most of those institutions and opinions which have been found common to Egypt with the southern nations of Asia were also discovered in the transatlantic kingdoms of Peru and Mexico. Some illustrations of this fact have already been given, but it may be interesting to furnish one or two additional proofs. The fourfold division of caste which is continued in India and Ceylon to the present day, which anciently prevailed in Assyria, in Persia, and in Egypt, and of which in the last country the traces are yet discoverable, was also noticed by the Spaniards as prevalent among the Peruvian and Mexican people. The same may be observed of the doctrine of the Metempsychosis or transmigration of souls, of the privileges of the established Priesthood, of the portion of soil allotted to their support, which in Egypt and in the American kingdoms was a third, and of various rights and customs as well political as religious. Architectural resemblances have I believe also been remarked, particularly the prevalence of the gigantesque style of decoration which is so remarkably developed in sculptured caverns of Elephanta and Ellora.[21] The existence of analogies like these is one of the most interesting subjects of speculation which the field of history presents and serves in some incidences to throw a clearer light on the origin and connection of nations than can be obtained from any other source. Perhaps the best example of the application of this method of historical inquiry will be found in President Goguet's parallel between Egypt and China.[22] In this way too may the study of those forms in which national character or religious prejudice has been embodied, be associated with general history.

In concluding this sketch I had intended to enter at some length into the question of the moral and social influences of polytheism over those nations among which it chiefly prevailed, but the length of time which I have already occupied warns me that my observations must be brief. There is a learned treatise on this subject by Professor Tholuck, of which a translation will be found in a volume of the Student's Cabinet Library, but it is a treatise which can by no means be recommended to general perusal. We will first consider the question and its political and social bearings.

It is an oft quoted remark of the historian of the *Decline and Fall of the Roman Empire* that the popular superstitions of the ancient world were regarded by the vulgar as equally true, by the philosopher as equally false, and by the magistrate as equally useful.[23] The last opinion can only be considered just insofar as the presence of a dominant superstition may have tended to prevent too close scrutiny into the affairs of government. If it be intended to convey the idea that the influence of a system of error was favorable to the peace and welfare of society, the position cannot I think be defended. It is an observation of Plutarch, himself an apologist for heathenism, that the tendency of that system was to urge men to atheism or to the foulest superstitions and vices. Now, the former of these results, atheism, has never been considered as conducive to social order, and the observance of the laws; and of the second it may be observed that it was on more than one location necessary for the Roman Senate to exert its power in order to check the evils which from this source alone were spreading through society and threatened to disorganize its very frame. To this cause Livy attributes the suppression of certain mysteries at Rome, which had been nightly celebrated by promiscuous crowds, under the veil of a secret association. Not only had their meetings partaken of certain features, which it is to be feared were common to most of the religious mysteries of Greece and Egypt, but they had been made a nursery of perjuries, conspiracies, and poisonings. The expulsion at a subsequent period of the priests of Isis, and the causes which led to it, offered another and instructive example of the social abuses which may shelter themselves under the wing of a depraved superstition. Indeed, if the tendencies of Heathenism on society can in any instance be regarded as conservative, we may safely pronounce them to be conservative only of those elements and conditions which most need to be replaced.

The moral influence of Polytheism, that is to say, its effects on the character of individual man, considered apart from its social and political tendencies, remain now to be considered. Properly speaking, this is a question of fact, and is to be decided by the testimony of history. To enter into the discussion of that evidence here would detain us long, and it will suffice to notice some general conclusions which such an examination warrants. What I have to advance on the subject will be comprised into observations.

The first is that under none of the forms which polytheism has ever assumed, or which it is conceivable that it might assume, is it capable of effecting any real or permanent improvement in the moral condition of the human race. The reason of this I believe to be twofold: first, the want of a pure code and uniform standards of morals, and secondly, the want of some grand central principle, like that of the unity of the Divine Nature, to which the uniformity of the standard and the authority of the code might be referred. Isolated precepts avail but little. Man, the poet Cowper has said, acting in society on his own responsibility is like a flower in its native bed, but too often, "Man associated and leagued with man / Like flowers selected from the rest, and bound, / And bundled close, to fill some crowded vase / Fades rapidly."[24]

Reverse this simile as to its general idea, and it will apply to the moral virtues. It is in association alone, and united in the strength of a principle, that they obtain their full growth and maturity; solitary, and cut off, they wither, or shoot out unhealthily. It is thus that while polytheism has often contained the scattered fragments of moral truth, derived, without question, from the unwritten dictates of natural conscience, it has never been able to give them their due force or effect, but in some instances has distorted them into an utterly opposite meaning. Of this fact I shall offer but one illustration. It is well known that the religious code of the Hindoo scriptures is characterized by remarkable tenderness for animals, founded on the doctrine of the transmigration of souls, and that in various parts of India, from a religious prejudice, the use of meat is prohibited. We read also of hospitals for sick animals, and of other institutions and practices founded on the same prejudice, which I do not choose to name. The natural result of the system we should imagine would be, the cultivation among nations of a general spirit of kindness and humanity, and unwillingness to inflict or to witness pain, and a benevolent desire to relieve it. Let us now hear the testimonies of some intelligent travelers on the subject which I have selected from a note in Mill's *British India*. The writer observes, "Although the killing of an animal of the ox kind is by all Hindoos considered as a kind of murder, I know no creature whose sufferings equal those of the labouring cattle of Hindustan."[25]

For the same purpose another writer observes, "The Gentoos, though they will not kill their meat cattle, make no scruple of working them to death. Neither are they less inhuman towards their sick, a woman being brought to die among the tombs in my sight."[26]

To these testimonies, which may easily be multiplied, we may add that before the introduction of British power, such an institution as a hospital for the sick and aged poor was unknown in India, although the hospitals to which I have alluded may boast of their antiquity. The natural effect of superstition on the character is very strikingly displayed. And the fanatic will permit himself to be devoured by vermin without attempting to molest them, while his fellow creature perishes by the wayside, unnoticed. A loathsome observance is made the standard of excellence, and the merit which it is supposed to confer releases from the common obligations of morality. Such is the complexion of fanaticism throughout the world.

The above examples serve also to introduce and in some measure to illustrate my second observation, which is that while polytheism in all its forms is powerless to give effect to the decisions of natural conscience, its proper tendencies are invariably of a demoralizing character. There have been some cases in which these tendencies have been met with and counteracted by causes operating from without. Such was the case in Republican Rome in her early and best days. There the spirit of military virtue for a time resisted the encroachments of a too certain corruption. In Greece too we cannot doubt that the humanizing influence of letters and of civilization would cause itself to be felt and that without such influence the popular religion, although it could scarcely have been more sensual, might yet have been less human and tolerant. It is in the great communities of southern Asia, where man stagnates in hereditary bondage, that we see the worst results of Heathenism developed, in superstitions which are at once fanatical and gloomy, at once licentious and cruel. To this condition must it ever gravitate, although circumstances may determine the nearness of its approach.

The one cause is to be sought for in that grand defect of polytheism, the want of an author of moral distinctions, and of a high, pure, and authoritative standard, by which every deflection from truth and purity maybe estimated and corrected.

There are two distinct species of evidence which we are accustomed to employ in the investigation of Natural Truth: the evidence of demonstration and the evidence of analogy.[1] To one or the other of these we have recourse in questions to which that direct evidence afforded by the senses, or by testimony, is inapplicable. The first species of evidence, that of demonstration, is of chief utility in the higher physical and mathematical sciences, its office being to reduce observed phenomena to the dominion of fixed laws, and again from fixed and ascertained laws to deduce the succession of phenomena. This induction of laws and derivation of phenomena has constituted in all the more important sciences the labour of many minds, and of successive ages. Thus in the science of Astronomy, it is to Newton that we owe the discovery of the law of Gravitation. It is to him in part, and to the Astronomers of France in the next age in a greater degree, that we must assign the honour of having traced that law to its remoter consequences. It will be seen that the entire process, whether of induction or derivation, rests on the general principle, that there is an appointed order of nature, that her sequences of cause and effect are invariable. Hence, were our means of analysis sufficiently powerful and comprehensive, it would be possible from a knowledge of the existing state of the material universe, to determine its condition at any future period. One of the questions which I have proposed to consider this evening—the question of the stability and permanency of the existing planetary system, is of this character, and to a certain extent admits of being answered by that species of evidence which I have endeavoured to describe. I do not however, when I come to speak of the question, design to offer any special account of the evidence on which it rests. Such an attempt would certainly fail of its object. I shall on that subject restrict myself to a simple exposition of results.[2]

The second species of evidence to which I have alluded, the evidence of analogy, is of familiar application in the common events of life. It takes not the high and absolute ground of demonstration but confines itself to such affirmations as these—that where we see a likeness of cause or of associated circumstances, we may with some probability infer a likeness of result; that from an imperfectly seen, or partially unfolded plan, we may form some conception of the finished performance. It is manifest that this species of evidence admits of degrees, and that the confidence with which we regard its conclusions will materially depend on the clearness of the observed analogies. It is on such evidence that the question which I shall first consider, i.e., whether the planets are inhabited, must be decided, and the question is one of probability, and does not involve the long and connected chain of reasoning in which demonstrative evidence usually consists—it is obvious that it belongs to all, when in possession of the necessary facts, to form a judgment for themselves.

In order then to prosecute the analogy, we shall first consider the general conditions to which life as here manifested, appears to be subject; secondly, the adaptations of the earth as the abode of life; and lastly the presumptions in favour of such adaptation on the part of the remaining orbs of the planetary system.

Now the most striking characteristic of life, whether animal or vegetable, as developed in this planet is its periodicity—the circumstance that it is marked by alternate periods of action and repose. The sleeping and waking life of animals is a great illustration of this fact. Many plants, it is observed, shut up their leaves at the approach of night, and extend them with the dawn, and it is perhaps true of all, that the functions which they exercise during the hours of darkness, are different from those which are carried out under the influence of light: the exceptions to this remark, if any, are so few, and as doubtful, that we are, I conceive, justified in assuming that this alternation of opposing states, this constant order of successive activity and remission, is a law of organized beings as manifested here—a necessity to which all the tribes of earth are subject.

Another characteristic of the life around us, and one that is closely connected with the general law we have just noticed, is, that the animated structure, passing through certain appointed stages of renovation and decay, tends finally to dissolution. The dominion of this law is universal. Such however may not be the condition of life in the abstract, it may be merely an accident of its present state.

We may, and often do, conceive of life as manifesting itself according to a different law, not needing repose, not subject to decay; but this is not the character of terrestrial life, nor could it be, so far as we can judge of the adaptations of our present abode. Let us consider what those adaptations are.

In the first place, the revolution of the earth on its axis, and the consequent succession of day and night, is an evident adaptation to that law of the animal constitution, by which it is subject to the necessity of repose. Again, the presence of a surrounding atmosphere seems to be clearly adapted to that condition of perpetual renovation and decay in the animal structure, to which I have called your attention. The earth might have been formed without an atmosphere, and the various tribes of animal and vegetable nature might have been created free from the need of that purifying and stimulating influence, which the air perpetually administers. But for wise ends it has been ordered otherwise, and what I wish to remark is, that between the two arrangements in question, there is a connection not certainly of cause and effect, but of mutual adaptation. I might here observe on several other points of relation, but those we have considered are the most important, and the others will be noticed in the sequel.

Now the first point of the analogy which I propose to trace is, that so far as we are able to judge, every planet of the system revolves on an axis, and is enveloped in an atmosphere. The sun revolves in 23 days, but to him of course, there is no vicissitude of light and darkness. He moves surrounded with perpetual day, environed at the same time by an atmosphere of immense extent. Mercury and the other planets in their order revolve, but in different periods: Venus and Mars in nearly the same as the earth, Jupiter in 10 hours, Saturn in about the same time; those periods making the respective lengths of their days. The moon and all the satellites revolve around their own axes in the same period in which they accomplish their revolutions around their primary planets. Thus the lunar day is of the duration of a month.

The existence of the planetary atmosphere is detected in various ways, sometimes by the appearance of twilight on the planet, or evening shade separating its dark and its perfectly illuminated hemisphere. Sometimes its presence is shown by the burden of clouds and vapours which it supports, and sometimes by peculiar optical phenomena which are exhibited when the planet surrounded by an atmosphere passes over a fixed star or satellite, or over the rim of the sun disk. By one or more of these methods, for all are not applicable in all cases, has the existence of an atmosphere been proved in every orb that is not either too small, or too distant for sufficiently accurate observation. Venus for example has a very extensive atmosphere. The twilight, which borders her sun lighted realms, has in certain positions of the planet been distinctly seen. Mars is surrounded by an atmosphere of greater density, and Jupiter and Saturn have each a like appendage. The atmosphere of the moon is very rare, and of small extent. Indeed it is probable that in various respects, the constitution of the satellites differs from that of the primary planets.

Another condition which appears necessary to the maintenance of animal and vegetable life on this planet is the presence of water. For the supply and distribution of this important element, it is requisite that the earth's surface should be indented by oceans, diversified by hill and dale, spread out into extended plains, and crossed by lofty mountains. The water systems of continents and large islands depend in a great measure on the positions and magnitudes of their mountain chains. It is to this cause that we must ascribe much of the difference which exists between the green savannahs of America and the African deserts, between the well-watered plains of Europe, and the thirsty wilds of Australia. Let us then consider the evidences which favour the existence of water, and of a diversified surface, on the sister planets. The facts which observation has presented bearing on this question are the following.

On the disc of the planet Mercury, dark greyish spots have been observed, but from its distance and proximity to the sun, their existence only is known. The presumption is, that they are mountains.

Venus has unquestionably mountain ranges of great extent, probably much greater than any existing on the earth's surface. These are found to be invariable in form and position. The edge of her enlightened portion, like that of the moon, is broken and irregular. The horns are sometimes seen to be rounded off, as if, says Prof. Nichol, by the shadow of a line of eminences.[3] Bright points are observed within the boundary of her dark region, the peaks undoubtedly of isolated and lofty mountains. Her atmosphere supports clouds, the true character of which is shown by their change of place and varying aspect. Of course it is only by telescopes of extraordinary power and excellence, that these phenomena are revealed to our gaze.

The surface of the planet Mars is divided into distinct regions, which are marked respectively by a reddish and a greenish tinge. The former are supposed to represent continents, the latter oceans. It is established, that the relative positions of these do not change—that their boundaries are invariable. Thus it has been found possible to delineate on a map the surface of the planet. In support of the opinion that Mars is possessed of water, or of some similar fluid, there is another fact of a very striking character, which I shall briefly explain.

We know that during the winter of our hemisphere, the North Pole of the earth is enveloped in frost and darkness. During that period, there is a very great accumulation of ice, partly in the Arctic seas, and partly in the larger rivers of Siberia and of Northern America. At the same time the frozen soil is covered to a great depth with snow, which indeed is nature's provision, affording to the dwarfed and stunted vegetation of those regions its only protection from the clemency of the season. With the coming in of the Spring, the Pole emerges into the light of the sun, the frost-bound deep is loosed, the snow disappears, and the ice is drifted by currents into warmer seas. Thus the limits of that wintery domain which girds the Pole are gradually contracted, until the revolution of the seasons restores it to its former dimensions.

Now there is reason to believe that this condition of things has an exact counterpart on the planet Mars. As either pole emerges from darkness, for Mars like the earth has its order and succession of seasons, it is observed to be surrounded with a brilliantly white covering, the dimensions of which are gradually contracted during the period in which it remains exposed to the action of the sun's rays, until it totally disappears. To what shall we attribute this remarkable phenomenon, if not to causes like those, which on our own planet are productive of similar effects? If the white circle in question is not of the nature of ice or snow, why does it manifest itself at the poles, and why is it impatient of the summer? On the other hand, if it does possess the character which we have supposed, how close, how intimate, is the analogy which binds the remote orb of Mars to the earth we tread.

Between Mars and Jupiter revolve the asteroids, four small planets which occupy the place of a larger one. Of their physical constitution nothing is known, unless we suppose with some that their origin is connected with that of meteoric stones, and thence infer a likeness of internal structure. Now aerolites do not, so far as I am aware, contain any elements which are not found as a constituent of our globe, though the combinations are different. Sometimes they consist of malleable iron, with a small portion of that rare metal nickel. Of this kind was that vast mass which fell near the river Genesei, and which the wandering Tartars regarded with superstitious dread. Captain Owen I think mentions a district in south Africa, over which a shower of meteoric copper is believed to have fallen.[4] That these bodies come to us from the planetary spheres, admits scarcely of any doubt, and whether they indeed originated in the event which gave birth to the asteroids or not, they to a certain extent afford a presumption, that the elementary constitution of the other planets is similar to our own.

The planet Jupiter, viewed through a telescope of moderate power, appears surrounded by a series of belts, or of shaded lines parallel to his equator, and separated from each other by broad white spaces. It is generally thought that these intervening spaces are clouds, and that the belts are portions of the dark body of the planet seen between them. As the appearances are among the most remarkable which observation has presented to us, I shall endeavour to explain their supposed origin.

In obedience to a well-known law, the great body of the atmosphere in the frigid and temperate zones of the earth has a tendency to move towards warmer regions, and a like tendency affects the waters of the oceans. As the two great currents approach the equator, they take a westerly direction; for as the velocity of the earth's rotation from west to east is greatest at the equator, and as the currents on their arrival there do not partake of the same motion in an equal degree, they—so to speak—fall behind their place, and produce a general current both of air and water in the opposite direction, i.e., from east to west. In the ocean the current is sometimes impeded, and its direction changed by the presence of continents and large islands. Thus the great easterly current of the Atlantic perpetually setting into the gulf of Mexico is turned aside into the North Atlantic and forms the well-known Gulf Stream. In the air however the general easterly current is less subject to this interference, and under the name of the trade winds prevails round the globe. Particular interruptions do indeed occur, as in the Indian Ocean, when the heated planes of the Hindustan produce—if I may be allowed the expression—a trade wind of their own. Such exceptions apart, it may be said, that an east wind prevails on either side of the equator throughout the year.

Now on the surface of Jupiter, the causes which here produce the trade winds operate with tenfold power. Notwithstanding his vastness, it is ascertained that he revolves in 10 hours. This prodigious velocity of revolution has even sensibly affected the form of the planet. I say sensibly, for while the forms of all the planets have been thus modified, it is on Jupiter and Saturn alone that the protuberant equator and flattened poles are apparent to the eye. Produced by the same cause, his trade winds must in like manner exceed ours in strength, and extend not merely over a strip on either side of his equator, but over a large portion of his surface. It is then [not] unreasonable to suppose that his atmosphere, burdened with clouds, should present the appearance we have noticed, and indicate by their arrangement the duration of his prevailing winds. It is a confirmation of this view, that his supposed clouds have been seen dispersed over the surface, and that dark spots are observed (probably mountains) in the belts, but never in the white spaces between them.

The same thing will apply to the belts of Saturn which differ from those of Jupiter only in that they are of a more uniform character and less subject to change.

I have mentioned the existence of dark spots on the planets supposed to represent mountains. Assuming that they are such, it is an interesting point of inquiry whether those causes which have upheaved the Alps and Andes of the globe have operated with like effect on others. Our own satellite affords us to a certain extent the means of answering this question.

We know that in some parts of our world the volcanic energy operates with unceasing activity. The manifestation of this we recognized in the earthquake, another in the burning mountain. We are assured too that it was by this agency that infinite wisdom saw fit to work when preparing an abode for man. It broke up the stony crust of the earth and piled out of its fragments the everlasting mountains. Extinct craters on almost every line of eminences betokened the manner of its origin. Now that particular form of the volcanic energy which we observe in the eruptive crater, manifests itself in two ways. Sometimes it acts along the line of disturbance. Of this we have an example in the Andes; another in that chain of volcanic islands which skirts the entire Eastern shore of Asia, and presents a line of burning summits, extending from the polar seas to the equator. Sometimes again it acts around a centre of disturbance. Now to whatever extent this agency has been developed on the earth, there can be no doubt that its action has been much more general on the moon. Viewing her through a powerful telescope, we imagine that an amount of volcanic power, equal to that which has been expended on the entire crust of the earth, had been shot up in her lesser orb, and bursting from its prison, and so riven and shattered her outward frame, and so deeply pitted and corrugated her surface, as to leave nothing to the eye of the spectator but a wilderness of rocks and craters; and that the causes we have supposed, the volcano, and that which we call earthquake, have indeed produced the effect, is rendered the more probable by this one circumstance: Amid all their scouring disorder the same general laws are recognized in the deposition of the lunar mountains and craters, as in the arrangement of our own. We see then the distinction to which I have adverted between lines and centres of disturbance. In some cases we find the two forms associated, as when a series of volcanic ridges radiate from a central crater. Many other clear analogies have been observed, of which I shall here only notice one. It is the property of all the great mountain chains on our planet, that on one side they present a steep and bold escarpment, on the other a gradual slope. Thus the mighty range of the Himmaleh [i.e., Himalayas] rises like a rocky barrier from the sheltered plains of India, but descends by a gentle declivity to the elevated steppes of Central Asia. Now the same property is said to belong to the lunar mountains. Taken in connexion with what I have remarked respecting the existence and arrangement of lunar craters, it seems to point out most clearly a similarity of the causes by which effects as similar have been evolved.

To the above statement it might be added that some observers assert with confidence that they have seen a bright light shining for a short period on the enlightened part of the moon's disk and gradually fading away. This has been supposed to be connected with a volcanic eruption. Others have imagined that they saw coruscations of electric light resembling the aurora borealis on the dark hemisphere of Venus, but further confirmation of the existence of this phenomenon is perhaps needed.

It will have been observed that many particulars of the general analogy which I have endeavoured to trace between the earth and the other planets do not apply to the moon. For want of water, the extreme rarity of her atmosphere, the total absence of those traces which the congregation of rational and intelligent beings on her surface might be supposed to have wrought, preclude us from affirming with any colour of probability that her orb is the seat of life. At the same time it may be proper to notice that some of her deeper and more sheltered valleys are said to reflect a green light which may for aught we know be the hue of vegetation. Considering how slight are the probabilities in favour of her being inhabited, it may be asked, why I have been so anxious to establish a relationship between her mountains and our own. I answer by remarking that on account of the immense distance of the primary planets we cannot hope to ascertain by observation to what cause the elevation of their mountain ranges is due. But if it can be shown that the process is the same in our satellites as it is here, there is good reason for supposing that the law is general and that it embraces in its operation the remaining planets of the system. Taking this view in connexion with the proved existence of water or of some similar fluid on their services, we see that their physical geology must in its principal features resemble our own. There is the same upheaving central force, the same abrading and dissolving agency. May we not extend the analogy and suppose that on the floors of their oceans are being deposited the strata destined to emerge in future continents and to preserve in enduring archives of stone the records of past changes?

We have now traced the general features of the analogy which was proposed to be considered, so far as observation has yet revealed them to us. In one or all of the planets, we have found evidence of the existence of clouds of water or of some fluid possessing similar properties, and of an atmosphere. Their surface, we have seen to be diversified with hills and valleys, and in one instance at least to present a supposed division of seas and continents: we have observed ice gathering around their poles, and currents like our trade winds manifesting themselves in the arrangement of their equatorial clouds. Their internal constitution we have reason to think similar to that of our own planet, though differing in respect of greater or lesser density; the agents of their physical changes are similar also. And now arises the question, to what final object does this analogy point? Are we justified in inferring that the planets were constructed to be habitations of life?

It is obvious that the inference supposed is the most natural one. If we discover on other planets such arrangements as on our own have a manifest adaptation to the wants of organized beings, it seems perfectly agreeable to our common modes of thinking to suppose that the planets are inhabited also. Let us examine whether there are grounds on which such a deduction may be formed as a logical and necessary consequence.

I think that the possibility of doing this depends on the admission that the universe is the work of an intelligent being, who acted not capriciously, but on fixed and immutable principles. Without this admission, I do not see how conclusions drawn from one part of the general system can be applied to another part. The ground of connection between those parts consists in this—that they have one common Author. And it is on this ground alone that we can maintain any deductions from the analogies which they present. Viewing the universe thus, in relation to its Author, we may from observation upon those parts of His works with which we are most familiar, derive some knowledge of the principles by which He has thought fit to direct the exercises of His creative power in reference to the whole. The conclusions which may with safety be drawn are, I conceive, the following:

From the fact that nature is governed by general laws we may infer that its Author acts by general rules, and hence, that in those parts of the universe which lie beyond the sphere of our observation, the same government of fixed laws is maintained. We have a remarkable illustration of this truth in the fact recently established that the law of gravitation extends to the fixed stars, where lesser suns revolve around greater, as the moon around the earth.

From external nature, and from innumerable examples in the animal and vegetable economy, we may infer that the Creator in the accomplishment of his designs, operates by secondary means, which are for the most part intelligible and uniform. The evidence of design and adaptation are indeed so palpable that the object of contrivance can in very few instances be mistaken. From the fossil remains of extinct animals, Cuvier inferred their forms and habits.[5] The confidence which we attach to the results of such inquiries rests on this—that we believe there are evidences of design even in the structure of a bone, or the convolutions of a shell, and that the object of the design may be inferred from the uniformity of general principles by which all the Creator's works are characterized.

From the present condition and past history of the earth, we may infer that it was specially designed by its Author to be an abode of life—not of life in mere subservience to the wants of man, but of life accomplishing in innumerable tribes of animated beings, its proper purpose of activity and enjoyment. The view which most faithfully represents nature, and most truly interprets the declarations of Holy Writ is, that there had been exceptional interpositions of creative power peopling the earth with successive races of beings adapted to its changing condition. Prodigious as is the development of the powers of life at the present time, it may be doubted whether it is equal in extent to that of remote eras. How exuberant must've been the vegetation from which the veins of coal were deposited! How great the abundance of animal life which covered with bones of elephants the inhospitable shores of Siberia, and keep them into vast accumulations on the islands of the polar sea!! But it is not only with respect to the larger forms of life that this prodigality of nature has been manifested. The discovery of the fossil infusoria of slate, establishing that extensive beds of that substance consist merely of the remains of animalcules, similar to those which are now found in stagnant waters, is a fact pregnant with instruction.[6] Again that mysterious law, by which the decay and dissolution of one animal structure supplies to countless others the means of existence, and death administers to life, shows in an extreme case how wide is the dominion given to the latter. It is not affirmed that the maintenance of life was the sole object for which the earth was called into being, but we are justified in asserting that such was at least the chief design of its Author.

Now if these conclusions are admitted, if it is granted that the Author of nature acts consistently in all his works, and that he accomplishes the purposes of his will by special means and adaptations, that his purpose with respect to the earth is, that it should be an abode of life, and that the mass and adaptations by which that purpose is accomplished are employed in other worlds, I see not how the conclusion is to be resisted, that those other worlds are intended to be habitations of life also. Refusing to admit the inference, we fall on such contradictions as these: that he acts inconsistently whose character is immutable; that he has created worlds without end or object, who made nothing in vain; that he has chosen to preside over an empire of death, who is the source and fountain of life and enjoyment.

I have observed that in maintaining that the planets are constructed to be abodes of life, we do not assert that this is the only object for which they were created. It is characteristic of the Creator's proceedings, to accomplish many purposes by one system of means. We may thus conceive that while each planet is to its own inhabitants an abode of life, it may be designed to answer important ends in respect of the other planets. The science of navigation which is of such great consequence rests almost exclusively on astronomical observation. Again, a planetary system is a mechanism so wonderful, that we may suppose one design of its formation to have been, to display not simply the vastness of creative power, but in a more eminent degree, the resources of creative wisdom. That refined and efficacious instrument of thought, the modern analysis, has chiefly arisen out of the study of the planetary arrangements. It may thus be imagined that the material universe has been adapted by its author not only to the physical but also to the intellectual constitution of his rational offspring. We might pursue these speculations and inquire whether there are not also examples of adaptations to the moral, as well as to the intellectual and physical, constitutions: as for example, whether there is not an analogy between the government of material nature by fixed laws, and that of moral agents by the unalterable rule of duty. I merely notice these as questions, which although they may not lead to definite conclusions, can hardly fail of affording interest and profit.

We have now considered the designs for which the planetary system was called into existence. It remains only to inquire whether it possesses the element of stability

There is reason to suppose that the planetary system has not always existed in its present state and form. It may once have been in the condition of the nebula, like perhaps to that which shines with a different and irregular splendour in the constellation Orion. Out of this formless and chaotic state may have gradually been evolved the order and harmonious arrangement which it now presents, a system of lesser orbs revolving around one predominant mass, in paths differing little from circles in the same direction, from west to east, and in nearly the same plane. The question to be considered is, whether this is the final condition of the system—whether, natural causes alone considered, such an order of things will always prevail. The results which the highest application of mathematics have yielded in reference to this question are the following:

First. One planet revolving around one central sun, through space, void of resistance, would constitute an undisturbed and permanent system. The planet would revolve in the same orbit, and with the same relations of distance and velocity forever.

Second. One or more planets revolving under the same conditions around a central sun, would constitute a disturbed but permanent system. The system would be a disturbed one, because each planet would, by virtue of its attraction, interfere with the motions of all the others. It would be a stable and permanent system, because all the irregularities of mutual disturbance would be confined within narrow limits. Increasing to a certain extent, they would afterwards diminish, and diminishing again, increase in such manner, that the system could never depart far from a mean state. If the orbit of the planet is at one period dilating, as is now the case with the earth's orbit, it will subsequently contract. If the inclination of the orbit's plane to the plane of the equator is diminishing, and such is the condition of the earth's now, such diminution will be followed by increase. The dilation of the earth's orbit, could it forever continue, would indefinitely lower the mean temperature of the year. In like manner, the other irregularity which I have named, would eventually abolish the distinction of seasons. Both have however their determined bounds which they cannot pass. All the perturbations of the planetary system produced by mutual attraction are of a self-correcting character. The unassisted intellect of man has achieved no higher triumph, than in establishing this great principle of the economy of the universe.

Third. Planets revolving around a central sun through a resisting medium cannot constitute a permanent system. The effect of the resistance in any given period may be but slight, perhaps imperceptible; accumulated through indefinite ages, it will cause each planet to revolve in a lessening orbit, and gradually to approach the sun. When once the existence of a resisting fluid in space is established, the doom of our own planetary system, and of all others, may be regarded as settled. Nothing short of a special interference with the laws of nature can avert its coming. The question then of the stability of our system ultimately rests on this: whether the planetary spaces are filled with a resisting ether or not. Reasoning from natural causes, we should say that if there is such an ether, the present arrangements of the universe exist but for a determinate period. If there is not, they may, unless other agencies interfere, endure to eternity.[7]

Now an opinion in favour of the existence of an ethereal fluid diffused through space has been very prevalent and it rests on two distinct kinds of evidence. It is thought to be proved by certain phenomena connected with the action of light, and also by observations on Encke's comet.[8]

With respect to the first species of evidence, it is now generally believed that light is produced by the vibrations of a highly elastic and refined ether, as sound is produced by the vibrations of the air. The facts on which this theory is founded are very numerous and consist of a great variety of phenomena observed under different circumstances, as when light passes through narrow apertures, or is transmitted through crystals and precious stones, such as Iceland spar, the amethyst, [and] the topaz. These results it is thought can only be explained on the supposition that there is an ether diffused through space, as vibrations, propagated with inconceivable rapidity, excite through the medium of our visual organs, the sensation of light.

Encke's comet is nothing more than an attenuated luminous vapour, revolving round the sun, in about three years and a half. If on any body, the effects of resistance could be perceived, it would be on one like this. Accordingly it is said that such an effect has been observed, and that the resisting ether is thereby rendered actively manifest. But it is to be remarked, that no effects of resistance have been traced in the motions of the planets. The disturbances produced by their mutual attraction were observed even before the cause was understood. They have been made the subjects of calculation, and enter into the construction of the most accurate tables. But the effects of resistance, constantly accumulating from the earliest period of accurate observation, are still imperceptible. This does not prove their non-resistance, but only shows that they are so slight, as to require the lapse of far longer periods before they can sensibly interfere with the existing arrangements of the universe.

From the above results I think we are entitled to infer that the planetary system may exist through periods of duration inconceivably great, but that it is not constructed for eternity. Indeed it is probable that there is no combination of things in the material universe to which eternal duration can with propriety be ascribed. We do in a certain sense speak of the eternity of the visible heavens, and we sometimes apply such modes of expression to describe the least perishable of the movements of human power, but it is only figuratively, and by comparison with the shortness of our own lives that this language can with any propriety be used. The physical agencies which under divine appointment have educed from a chaotic state the shining order of the heavenly frame would alone suffice for its dissolution.

We have however no right to affirm that the dissolution of the system will be left to the slow though certain operation of the physical causes which we have considered. The period which they would require is so incalculably great that long ere it shall have elapsed, all the designs for which the planetary worlds were called into existence may have received their final accomplishment. The power which has interposed to create, may be interposed again to destroy, and to create anew. The traditionary belief in all ages has pointed to a destruction by fire of the present order of things. This was the doctrine of some of the ancient philosophical sects in Greece, and it is finally alluded to by Ovid.[9]

An authority higher than that of Grecian sage or Roman poet has told us [of an]other day "in the which the heavens shall pass away with a great noise and the elements shall melt with fervent heat."[10] We should err if we should suppose that a declaration like the above is given us for the purpose of gratifying an idle curiosity as to the manner in which the great designs of Providence are to be realized, but we may lawfully receive it as an intimation that the existing constitution of material things is to pass away and to give place to a higher. The discussion of such questions does not however fall within the province of natural philosophy. It is well for us to know where that study should begin and when it should end. One of the best lessons it can teach us is, to trust, in things unknown, the wisdom and benevolence which are conspicuous in the things that are known.

In the reports of the progress of the institution which have been annually prepared for your information, you have been accustomed to hear its prosperity ascribed to the rule by which works on party politics and controversial divinity are directed to be excluded from the library.[1] You have not, perhaps, been told that this rule has never been systematically obeyed—that its existence has been from the very foundation of the institution a cause of perplexity and dissension in the committee, and that it has at length brought your institution into serious difficulties. Will you allow me for a few minutes to draw your attention to these facts? I am anxious to do this because I believe them to be important and because I think that the question which they suggest is one which can no longer be evaded. In the first place, I remark that the rule has never been systematically obeyed. Its strict observance is indeed impracticable. We require a previous definition of the term party politics and controversial divinity. The authorities of the institution have never agreed upon this point. Different committees have assigned to them very different meanings, and even the same committee has given to them at one time a lax [and] at another time a rigid interpretation. In the stricter and only definite acceptation of the terms, every work is a party work as far as it maintains the opinions of a party, [and] every work is controversial which contains a defence or decided expression of opinions that are controverted. As this view has been often acted upon we may for the present regard it as the true one. Now consider for a moment to how great an extent politics and theology have influenced our national literature. They run through our history and much of our poetry and philosophy—they enter into our best works of fiction—they tinge our magazine and periodical essays—they are the two poles to which every stronger mind is irresistibly drawn.

Rigidly interpret the rule for their exclusion and you shut up your library against the ablest writers, the most earnest thinkers, of this and every age. You banish from your shelves Locke and Smith, Taylor and Barrow, Chalmers and Whately, Alison and Vaughan.[2] How will you supply the place of this class of writers? What provision can you make for the instruction of the old, for the moral nurture of the young, if you forbid all inquiry into the greatest and most serious questions with which a human being is concerned? The great extent to which the practice of novel reading prevails among the younger members of the institution ought to be a subject of deep concern to its friends, and should stimulate them to inquire into the real merits of that system of exclusion under which so pernicious a habit has been permitted to grow up.

In the second place the impossibility of enforcing the rigid observance of the rule has led to innumerable dissensions in your committee. This will not appear surprising if you consider that when obedience to the strict requirements of the rule is impossible the question must arise, to what extent and in favour of what interest it may be violated. From the bias of prejudice the best are not free. Hence unconscious aggression on the feelings and opinions of others, real or supposed injustice, [and] angry recrimination. Under your first president such dissensions are said to have been frequent and high, and they have not entirely ceased under his present able successor.

A temporary difficulty into which the Institution has been brought by the operation of this Rule, is now to be noticed. A few weeks since, the Secretary received an offer, made on behalf of a Lady, of a donation of books which were accepted by the Committee. When received, it was found that most of them contravened the Rule. It is true that they were works of the highest character, written in the most tolerant and most Christian spirit, bearing on the title pages the honoured names of Whately and Sumner, with others almost equally eminent.[3] They treated controversially of politics and of the Christian Religion, and they were therefore deemed indefensible. The Secretary was accordingly directed to follow his letter of acceptance by a second, requesting permission to return a large portion of the gift. A short time previously, a donation of the works of Dr. Arnold had been accepted.[4] It will not be contended that none of these are controversial. I use the word in no disparaging sense. It will scarcely be disputed that some of them are eminently controversial. Into such inconsistencies are your Committee betrayed, by the differing interpretations of a vague rule.

But there is, beside this, a standing difficulty under which every conscientious member of the Committee must labour as long as the Rule is suffered to remain. It is the difficulty of reconciling Conscience with Expediency. While the Rule exists, it demands obedience. If that obedience is strictly rendered, the interests, the very utility of the Institution, are sacrificed. Hence a conflict of feeling, more painful than the conflict of opinion. Herein is a plain wrong. The highest advantage of the Institution is dearly purchased by the sacrifice of a single conviction of duty on the art of its meanest servant.

And for what purpose is a Rule so fraught with evil maintained? To prevent discord? Has it not rather tended to promote discord? Is there evidence in the experience of other Institutions that the admission of standard works on the forbidden subject has produced disunion. The inquiries which I have had the opportunity of making on this subject have led me to the conclusion that no such danger is to be apprehended. Society is indeed founded on mutual concession and forbearance; they are part of the lesson of life. Should it then be proclaimed that a literary institution is the only place in which their operation cannot be trusted? Whenever angry discussions with reference to politics and theology do arise, it is at least evident that it is not from the perusal of the highest class of works on those subjects. The patient labour of analysis, and the calm deductions of reason, are not the means by which party spirit achieves its miserable triumphs.

There are some subjects in which Indifference is Ignorance or Weakness. Such are all which relate to questions of Human Duty. The Christian religion is the foremost of these, and I hold that the investigation of its principles is a proper study for the Members of a Mechanics' Institute. The diversity of the forms and modes in which it presents itself to different minds, affords us reason for its exclusion, but should rather instruct us to be impartial in our choice of authors. I dislike Puseyism but I would permit it to speak for itself in Newman on Development.[5] I am not an Unitarian but I would make room for the mild and eloquent Channing.[6] I have no moral sympathies with what is called Calvinism but would find a place for the unrivalled dissertations of Jonathan Edwards. And the same liberal system would extend to Politics and Social Economy.

From the doors of your Library every jealous barrier should be withdrawn. On its silent shelves the Free Trader and the Protectionist, the feudalist and the advocate of change, should array their facts and marshal their arguments, for the effectual battle of Truth. From a selection fairly made, from the different schools into which opinion is divided, and in such measure as the demand might warrant, I should apprehend no increase of a factious and intolerant spirit. On the contrary it is among those who have read and thought the least, that the unhappy tendencies of party and of sect chiefly prevail.

The dictates of a wise and liberal policy require, that you make your Institution as free and open as possible, that you record it as the conviction of your minds, and the fundamental principles of your Society, that there is no branch of learning, human or divine, for the study of which facilities may not be provided in a Mechanics' Institute.

In lectures and in classes, I have not much faith. In a well-organized and impartially conducted Library open to every class of readers and of subjects I have great faith. The right to acquire knowledge like every other right involves a responsibility. It may be neglected or abused, and for such neglect or abuse you will be answerable, but the right exists, and they are at best mistaken in their friendship who would deprive you of its exercise. Choose your representatives in the Committee with care, and having chosen them, confide to them in a free and generous spirit, the interests of your Institution, abolishing all vexatious restraints upon the liberty of their decisions. The only restriction which is worthy of the Institution is that imposed by the good sense and right feeling of its members. As this is sufficient to prevent the introduction of immoral and licentious publications, so it would bar the admission of all work in either Politics or Religion, which should not be conceived in a spirit befitting the pursuit of Truth, and answering to the dignity of the subject.

I address myself this evening to the young men of Lincoln, and more especially to that portion of them who have been benefited by the late movement for the early closing of shops.[1] Many of you who are here present, have probably taken a part in this movement, and I feel that I cannot better preface my observations, than by congratulating you on the success which has attended your efforts. On the right use of that leisure and opportunity which you have thus obtained, many happy results may be made to depend. I offer to you my congratulations, not simply because I set a high value upon the advantages which are secured to you, but because I think, that the very fact of your having obtained them, may be regarded as a proof that you have in some measure deserved them. For, in concessions of this nature, something more than justice is usually considered; and I doubt not, that in the minds of those who have listened to your demands, a sense of the equity of your claims was associated with the conviction that they might with safety be conceded—that the rights for which you asked would not be abused. Such a conviction must have been founded, not less in experience of the past, than in hopeful views of the future. I think therefore, that the so far successful issue of your temperate but persevering efforts, reflects a certain measure of credit upon yourselves, as evidencing that in the opinion of your employers, and of the public generally, you were prepared for the right application of the boon which you were about to receive.

I have spoken of the advantages of leisure, and opportunity for improvement, as of a right to which you were entitled. I must now remind you, that every right involves a responsibility. The greater our freedom from external restrictions, the more do we become the rightful subjects of the moral law within us. The less our accountability to man, the greater our accountability to a higher power. Such a thing as irresponsible right has no existence in this world. Even in the formation of opinion, which is of all things the freest from human control, and for which something like irresponsible right has been claimed, we are deeply answerable for the use we make of our reason, our means of information, and our various opportunities of arriving at a correct judgment. It is true, that so long as we observe the established rules of society, we are not to be called upon before any human court to answer for the application of our leisure; but so much the more are we bound by a higher than human law, to redeem to the full our opportunities. The application of this general truth to the circumstances of your present position is obvious. A limited portion of leisure in the evening of each day is allotted to you, and it is incumbent upon you to consider, how you may best employ it. I propose in this lecture, to offer some suggestions which may assist you in the consideration of this important question.

I wish, in the first instance, to call your attention to that wise arrangement of Providence, by which there exist at once, so great a diversity in the human mind, and a wide variety of objects in which it may innocently seek for gratification. There are so many things which it is good to learn, so many which it is good to do, as to afford employment for all our faculties both of thought and action. However various our tastes, our dispositions, or our powers, we have before us an ample field for their exercise. Reflect for a moment on the wealth of ideas which is bequeathed to us in the writings of our poets; on the stores of instruction which are opened in our histories; on the lessons of wisdom contained in the recorded lives of those who have adorned our race. If we desire to acquaint ourselves with the structure of the Universe, how many subjects of inquiry and of meditation present themselves! What wonders in the heavens; what a glory and beauty in the world around us! What indications of order and presiding intelligence throughout the whole! Dr. Chalmers has remarked, that every investigator of Nature is led to regard his own particular department of research as the richest in interest and wonder. How replete and gorgeous, he then observes, should we consider the whole to be![2] If we quit the study of material Nature, and apply ourselves to the pursuit of Truth in the province of moral and social inquiry, another Universe, not less replenished with wonder and with interest, presents itself. The rise and fall of extinct nations; the mission they accomplished in the world, and the work which they left to be done by us; the causes on which national wealth and virtue and happiness depend; these are questions which are second to none within the range of the human intellect, in dignity and interest. In short, to whatever province of the kingdom of Thought we turn our attention, we find abundant scope and reward for the activity of our inquiry.

Nor is the sphere of our activity to be confined to the exercise of the intellectual powers. The body has its claims also. An ancient poet, when he had summed up the diversified objects of human desire, concluded with a simple prayer for a sound mind in a sound body.[3] A healthy frame is indeed in some measure necessary to the possession of a vigorous mind. Happily we may by the same course of action consult the advantage of both. The exercise which imparts vigour to the bodily system, may at the same time furnish the mind with agreeable ideas. Nor are athletic sports and games to be looked upon as merely conducive to health and recreation. They may assist in producing a vigorous and manly character of mind, all in encouraging a free, generous, and open disposition. Thus a most varied store of innocent enjoyments is accessible to those who know how to avail themselves of it. I shall return to this subject in a future part of the present address.

The position which in the preceding remarks I have been anxious to establish, is, that however varied your tastes may be, whether you are disposed to seek gratification in books, or in active exercise and free intercourse with nature—there is a provision of virtuous enjoyments, sufficient to meet your wishes and to reward your efforts. If you seek gratification in those pursuits from which virtue turns aside, you do so without excuse. The constitution of things has not been so ordered, as to render it necessary that you should have recourse to such means. I say nothing here of the fading character of all merely sensual pleasures, and of the ruin and self-reproach which their excess entails, because all experience shows that the most effectual teaching is that which appeals to our better principles, and which seeks rather to attract than to warn. I content myself by remarking, that you may turn your leisure to the best account, as respects not only your personal improvement, but also your individual happiness, in a manner which is perfectly consistent with the requirements of Duty and of Virtue.

Presuming that your assent is given to the remarks which have now been made, I shall proceed to offer a few suggestions, with reference— first, to the education which you may derive from books; and, secondly, to that which may be obtained from other sources. And, first, of the education which may be derived from books. There are none of the monuments of past ages, in which so much of the accumulated results of human labour is presented to us, as in books.

Considered merely as a memorial of the industry of man, every well-filled library is more replete with wonder, than are the most costly of material structures. The most renowned of ancient cities, could we repair the ruins of time, and restore its lost magnificence, would scarcely represent a larger aggregate of toil, than the single library of the British Museum. Such a collection excels every other result of labour, not less in character than in extent. It presents us with the issues of all past time; it unfolds to us the great discoveries of science; it brings us into acquaintanceship with men who were, intellectually and morally, of a larger than the common stature of our race. We should esteem it as a high privilege, to be admitted to this intimate fellowship with the wise and eminent, not of this place or of the present time alone, but of all times and generations. Something of this feeling has been expressed by the poet Southey, in those beautiful lines on his Library—

> My days amid the dead are past;
> Around me I behold,
> Where'er these casual eyes are cast,
> The mighty minds of old.
> My never-failing friends are they,
> With whom I converse, day by day.[4]

And the feeling which the poet has thus expressed, must be common to all with whom books have been familiar. How many have found in them a refuge from the cares and distractions of life! How many have derived from them the lessons which have enabled them to bear themselves with upright courage, and serene integrity, in the most trying and difficult circumstances! It is our lot to live in an age in which they are cheap and abundant, beyond all precedent. Our only difficulty is in the selection. We should manifest a proper sense of the inestimable privilege of our position, and study to use it aright. Let us more particularly consider how this may be done.

One of the most important studies in which you can engage is that of History. The manner in which you should proceed, will depend very much upon the particular object which you propose to yourself. It is important for you to determine, whether you will endeavour to obtain a general acquaintance with this branch of learning, or whether you will more thoroughly investigate the history of some particular period or people. To those who have but little leisure at command, the latter appears to be the preferable course. I will enumerate some of the principal sources of information which are open to you.

On the subject of Universal History, not to mention the larger compilations (some of which are original, and some translated from the French and German), you may look with advantage into Tytler's *Elements of General History*, a very elegant work which may now be procured in a cheap form, Heeren's manual of Ancient History, and Dr. Cooke Taylor's Students' manuals of Ancient and Modern History, published in two small octavo volumes.[5] The latter work, or Tytler's (already mentioned), would in all ordinary cases be sufficient; and they are moderate both in price and size.

Among the vast multitude of works which record the histories of particular nations, I need only mention, and this rather by way of illustration than as recommending to you any definite course of reading, Thirlwall's Greece, Sismondi's Histories of Rome and of the Italian Republics, Robertson's America and Scotland, and various well-known histories of England, from Hume to the *Pictorial*, together with the miscellaneous volumes published in Lardner's *Cyclopaedia*.[6] To these I might add, if I were speaking to those who have much leisure at command, Niebuhr's *History of Rome*.[7]

Another and very important class of historical works is confined to the elucidation of particular epochs; and I know of none that I should more readily commend to your notice. Robertson's *History of Charles V*, Schiller's *Thirty Years' War*, Clarendon's *History of the Great Rebellion*, with Forster's *British Statesmen*, Voltaire's *Age of Louis XIV*, and D'Aubigné's *History of the Reformation*, are all of this class.[8] Such works are usually more interesting than general history, because they abound in personal detail and delineation of character, and unveil to us those secret springs of action, or silent workings of opinion, from which revolution and historic change have resulted. It is to be borne in mind, that this species of history has commonly been written by men, who have held strong and decided views on controverted points. We should take this into account, in forming an estimate of the value of their conclusions; and should in general endeavour to hear both sides of every disputed question.

Further, there is a class of works, the object of which is to unfold the philosophy of history. Of this class, are Guizot's *History of European Civilization*, Taylor's *History of Modern Civilization*, Vaughan's *Age of Great Cities*; and many others.[9] The design of such works is, to determine the influence of laws, of opinions, of the essential characters of races, and of other circumstances, in modifying the development of a people's history; to strike the balance between different eras, and so settle the general question of progress; and to solve other problems of a like nature. I need scarcely mention, that a very considerable amount of historical learning is presupposed in the student who would engage in such inquiries.

Lastly, we have the Biographies of eminent men, who, though not filling any important place in the general history of their times, have left behind them the enduring memory of genius or virtue. Of all the forms of history, this is the most attractive; of all modes of teaching, the most impressive and effectual. To think that we can stand with Newton while he resolves with his prism the beam of light; that we can accompany Howard on his mission of benevolence, and survey in kindred sympathy with his, the moral ruins of mankind; that, even from the most profligate page of our country's history, we can turn aside to hold communion with the lofty Sydney or the pure-minded Leighton; this is no small privilege.[10] To the young, especially, is the study of biography important. The high standard of character to which they may thus be familiarized, may produce the happiest effects on their future lives.

Allow me now to offer a few suggestions as to the mode in which History should be studied.

It is scarcely necessary for me to tell you, that you should make it your first endeavour to understand the meaning of what you read. This is a general direction, applicable to every study. Unhappily, it is often slighted; and too many, through haste or indolence, neglect the obvious condition on which all solid improvement depends. They who require the use of a dictionary, should not be ashamed to have recourse to one.

It is of great importance to be provided with a good atlas, and trace the different localities to which, in the course of the narration, we are successively led. This direction is important for two reasons. The first is, because the method it enjoins, is essential to a right understanding of History. The progress of events, especially in time of war, depends, very materially, upon the physical character of the region in which the scene of action is laid. The extraordinary struggle between Philip II and the Protestant Netherlands, was throughout modified by physical circumstances. We never should have had, side by side, the brilliant but mournful pages which record the sieges of Leyden and Antwerp, if the natural features of the region in which those cities are situate[d], had been other than they are. Again, the successive attempts of Austria, France, Burgundy, and Savoy, to subjugate Switzerland, might have had a very different result, had it not been for the mountain barriers, and, we may add, the spirit of mountain freedom, by which that little country was defended. The other reason, why the use of maps should accompany the study of History, is, because they afford a sensible link of association, by which the events recorded are more securely retained in the memory.

In the choice of subjects for more particular attention, give the preference to those which either exhibit the struggle of great opposing principles, or the slow results of moral causes, in the gradual rise and decline of nations. If you choose the former, endeavour to understand the nature of the principles at issue, and, as far as possible, realize to yourselves the positions and feelings of the contending parties. From their excesses you may learn moderation,—from their courage and constancy, fortitude and faith. If you choose the latter class of subjects, those, I mean, which display the results of moral causes in the affairs of nations, be slow in forming opinions, and avoid hasty and incautious generalizations. To give an example of what I mean by this class of questions, I would instance that which proposes to estimate the influence of domestic slavery, in accelerating the downfall of the Roman Empire.

I have dwelt at some length on the study of History because I regard it as an extremely important branch of self-education, presenting sufficient variety to meet the tastes of all classes of readers, and requiring little preliminary preparation. I proceed, in the next place, to consider the claims of the Physical Sciences upon your attention.

That we may better comprehend the position, which the study of physical science should occupy in the general system of mental culture, let us briefly consider what physical science is, and contemplate the circumstances which attend its origin and its progress.

Every physical science commences with the observation of facts. These may either be common or familiar, as the tides, the rising and setting of the heavenly bodies, the course of winds in the air, and currents in the sea; or they may present themselves to our notice only under particular circumstances, and in virtue of special arrangements; as the attraction or repulsion of the wires in a galvanic circuit, and their influence on the magnetic needle, phenomena of which we avail ourselves in the electric telegraph. But under whatever circumstances arising, facts and observations are the only basis of positive science.

The next step in the chain of discovery, consists in the determination of the laws, by which the sequence of phenomena is governed. This is the business, partly of observation, and partly of reason. Thus it is found, to adhere still to our former illustrations, that the flowing of the tides, and the rising, culminating, and setting of the stars, do not follow in an arbitrary and capricious manner, but according to certain fixed and invariable laws of succession; and that, even where apparent irregularities are observed, those irregularities are themselves subject to rule, and are only the exponents of a higher law. A long series of observations has in each instance been necessary to establish the existence of laws, such as we have alluded to; and a still longer series to determine their precise form and expression. Let it be remembered that all this has nothing to do with the question, how this order of phenomena is brought about—with the question of cause. All that we are thus far concerned with, is the law, deduced from observation and comparison, by which the particular succession of events which we may have to consider, appears to be regulated. This constitutes the second great step in the progress of science; and we see that it calls into action those faculties of the mind by which it judges, and classifies, and compares; and in a less degree, the higher power of generalization.

The last and highest step in the progress of a science, is the reducing of the secondary laws of the phenomena to some higher and more comprehensive law, standing to them in the relation of a mechanical cause, and constituting a definite limit to our inquiry. In the case which I have already adduced, of the flux and reflux of the sea, and the apparent motions of the planets, the mechanical cause of this great succession of phenomena, is the force of Gravitation; of whose simple and primary law of action, all the more complex and varied laws, to which we have referred, are necessary results. In the present state of knowledge, it is the science of Astronomy alone that has fully passed into this third and final stage of its development. Other sciences are usually in the second, or, at most, in a transition state. We know the laws of the phenomena, but we have not fully determined the agency to which they owe their birth.

That process of the mind by which we are enabled to discover general truths, is called Induction. It is the highest exercise of our intellectual powers; and it is that which most specially distinguishes the mind of man, from that of the lower creatures. When, through its operation, a science has been brought to that state, in which Astronomy now is, a reverse process, to which we give the name of Deduction, becomes possible. We have already proved the existence of a cause, and determined the mode of its agency, from the observation of phenomena. We now take our starting point from the cause itself, and deduce from it, new phenomena unobserved before. That recent prediction by Le Verrier and Adams, of a new planet, with the fame of which the scientific world is still ringing, was a discovery of this class.[11] The disturbances which its unseen presence occasioned, were noted, and the law of gravitation then sufficed to determine its position and its path; and observation has fully verified the prediction of theory.

From the above sketch we may perceive, that the study of the physical sciences affords a varied exercise to the intellectual powers; that it teaches us admirable lessons in the art of investigating truth. Indeed, it cannot be disputed, that the progress of moral and social science, and the sounder views which are entertained on these subjects in the present day, are indirectly attributable, in a great measure, to the methods which physical science has taught. These are its incidental, but not its least important services. Its main design is the discovery of the laws and constitution of the Universe; and from the accomplishment of this object, it must ever derive its chief value. Of the services which it has rendered to the arts of life, it is not necessary here to speak.

It has been the fate of physical science, to be invested by one class of writers, with attributes which it never claimed; and to be treated by another, with undeserved neglect or contempt. At the time when Mechanics Institutions were established, the study of it seems to have been regarded as fitted to supply some great popular want, and to effect a species of intellectual regeneration in society. Such anticipations have not been realized; and it is easy to see that they were founded in a misconception of human nature. Man must ever be the great study of man. On the other hand, its claims are not unfrequently met by the sneer of flippant ignorance, or the frown of bigotry. It must be confessed, however, that it meets with such treatment, only from those who have the greatest need of its intellectual discipline. The degree in which it ought to be made an object of pursuit, must depend upon taste, leisure, and ability. Its greater discoveries should be known by all. Its more recondite mysteries, may be left to the few. In all cases, it should be looked upon as a part, and as only a part, of the education of a human being.

There is a view of physical science, which I hold to be not more ennobling from the dignity and grandeur with which it invests the subject, than it is worthy of our acceptance, from its literal truth and faithfulness. It is when we consider it in relation to its great Author and Founder. Viewed in this light, the laws of the material Universe are the expressions of his Will; their unfailing certainty is the symbol of His Immutableness; their wondrous adaptation is an argument of His Wisdom; their wide dominion a testimony of His Universal Presence. Regard the truths of Science in this aspect; and derision and scorn can find no place. We cannot, in any absolute sense, regard, as little or contemptible, the study of those means which Infinite Intelligence has not disdained to employ.

Treatises on Natural Philosophy are so numerous, that it would be difficult to give an account even of the best. Bird's *Natural Philosophy*; Dr. Young's *Lectures*, in the recent edition of Professor Kelland and Peschell's *Elements of Physics*, a translation from the German, give the latest exposition of principles, and may safely be recommended.[12]

We proceed, in the next place, to consider the claims which the Moral Sciences possess upon your attention. The term we use, is sometimes employed to designate not only Moral Philosophy, but also the general philosophy of the mind. However, we shall here confine ourselves to its stricter meaning, and shall consider it as simply denoting the science of duty.

Someone will here perhaps ask what need have we of a science of human duty. Are not the distinctions between right and wrong, sufficiently broad and distinct? Why should we seek to encumber with the dead forms of a science, the spontaneous convictions of the mind and feelings of the heart?

To this it is answered, that the general rules of right are, indeed, clear and unmistakable, but that in the complexity of human affairs, their individual application is often difficult and obscure. They who never make their duty their study, may with the best intentions fall into error. Especially in cases of controverted right, the bias of personal interest may become too strong for the unexercised judgment; so that while they seem to themselves to contend for what is right alone, they may, to others, appear to be guilty of flagrant wrong. And, beside this, the restless activity of the human mind, its eagerness to know, impels it to inquire, not only into the rules of moral action, but into their origin, their nature, their limits, and their final aim. Hence the science of Moral Philosophy.

And there is no study which has so deeply engaged the attention of the wise, in every age, as this. Solon's celebrated apothegm, Know Thyself, is only a recognition of the pre-eminence of its claims. It was the chief subject of the philosophical treatises of the ancient Greek and Roman writers. It formed the material on which Aquinas and the schoolmen, expended much of their dialectic subtlety. In the writings of the divines of the seventeenth century, it fills folios under the titles of Casuistry, Practical Theology, Cases of Conscience, etc. In the present day it is more usually presented in short systematic treatises. Wide as is the attention which the subject has received, it must be confessed that it is worthy of it all. If there is any topic which it is important for man to study, it is his duty. The labours of those who wrote folios upon the subject, into which no one now looks, were not misspent. They are to be regarded as a testimony to the moral nature of man—as a proof, that it was in the contemplation of Duty and of Virtue, that he was to find a meet exercise for his speculative faculties; as it was in their practical application, that he was to exert his active powers. The two finest things which I know, said a great philosopher, are the starry heavens above our head, and the consciousness of virtue within the breast.

Of modern systems of moral philosophy, that of Paley is the best known. Like all Paley's works it is clear in its details, and admirable in its expositions; but its fundamental principles may be regarded as lax and defective. Wayland's *Elements of Moral Science* is an excellent work; free from the objections which may be urged against Paley's, and well-deserving of attentive perusal; the author is President of one of the American colleges.[13] Brown's Lectures on Moral Philosophy, an edition of which has recently been published by Dr. Chalmers, are a very valuable contribution to the science. Sir James Mackintosh's *Dissertation on the Progress of Ethical Philosophy*, gives an interesting historical sketch of the subject. Dr. Whewell's recent work on the principles of Morality and Polity, should also be mentioned.[14] To the above we may add a treatise on Moral Philosophy by Jonathan Dymond, a member of the Society of Friends.[15]

The diversity of writers and of systems which we meet with in this field of inquiry, is no argument in favour of a neglect of its claims; for, in general, writers do not differ as to the rules of morals, or as to the practical application of those rules to the conduct of life, but only as to their origin and first principle. One writer places the foundation of virtue in a tendency to promote happiness, another in the fitness of things, another in obedience to the will of God, another in the indications of a moral sense within us, [and] another in the essential and underived attributes of the Divine Nature, a view which is very ably developed in the writings of Chalmers and Wardlaw.[16] Whichsoever of these views we assume to be true, considerations founded on the others may greatly assist us in our practical determinations; and thus, in all that concerns the necessities and utilities of life and conduct, no material discordance of opinion can arise. The practical conclusions of morals are in a great measure independent of any theory of their origin.

Undoubtedly there is a danger of a speculative excess, and of an over curious nicety and refinement in the study of morals. But it is a danger to which the present age is less exposed than, perhaps, any former one. The tendencies of these times are rather to the opposite extreme, and we are, it may be, too impatient of things of which we do not see the immediate practical bearing. Still, it should be remembered, that there is a contrary danger,—a possibility, actually realized in some cases, of forgetting, in a too minute attention to particulars, that the chief end and consummation of knowledge is practice.

We ought indeed to seek Truth for its own sake; and we cannot set too high an absolute value upon either rectitude of opinion, or consistency and accuracy of judgment. But it is our business to act as well as to know; and these two faculties of our nature, the speculative and the practical, so far from interfering with each other, may contribute mutual strength and support. If right judgments are necessary to rectitude of conduct, the converse proposition is true also. Rectitude of intention, and an earnest desire to carry into practice the truth to which we have already attained, are in some measure necessary to correctness of judgment. For belief is not altogether involuntary; but while it acts it is in turn acted upon by the habits, the feelings, and the will.

Before quitting this part of my subject, I ought, perhaps, to notice one or two objections which are sometimes urged, but more often silently felt, against the cultivation of the mind by the acquisition of knowledge. The most popular of these are, that education fosters pride, and that it generates a distaste for the ordinary business of life.

I will not deny that it is possible for an individual to be proud of his attainments;—of his real or supposed intellectual superiority. For of what indeed may not a man be proud? Knowledge is not the only thing on which pride may build. A man may be proud of his wealth, of his rank and station in society, of his physical endowments. He may be proud of his virtues; of justice, or benevolence, or truth; and he may be proud of his vices, of his moral deformity, or of his ignorance. He may boast of that which is good and honourable, and he may also glory in his shame.

In all these cases, we must be careful to distinguish between the origin of the sentiment and the object to which it is directed. We are not to suppose, that because a man is proud of some possession or endowment, that particular possession or endowment is the cause of his pride. Undoubtedly the sentiment originates in our nature. Is wealth the object of pride? You will not destroy that pride by the removal of the object, but only change its direction; and so likewise of knowledge, were it possible that by any accident of fortune it could be taken away. And of this possession in particular it may be observed, that although it may be associated with pride, its natural and most graceful alliance is with humility. The lives of men eminent for their attainments, give testimony to this truth. We need not speak of Newton, and Locke, of Adam Clarke, and Sir W. Jones.[17]

Similar remarks may be applied to the other ground of objection which I have named. A studious person might neglect his business for the sake of books; but if he does this, it is not his books that are to blame, but his want of principle or of firmness.

Further, there are in the minds of some, objections, rather felt than acknowledged, against the pursuit of learning, which appear to be founded on a mistaken apprehension of the meaning of certain passages of Scripture. Such passages, I mean, as those which speak of a "knowledge that puffeth up"; of "philosophy and vain deceit"; of "science falsely so called." Many of my auditors do not need to be told that those passages have reference to different forms of the Gnostic philosophy—a system of delusive speculation, exerting indeed a baleful influence upon morals and belief in its day, but long since exploded and well nigh forgotten.[18] It was a system, indeed, quite as much of false religion as of false philosophy. With many vain and presumptuous notions on the subject of Creation, of the origin of evil, and of questions similar to these, it inculcated practices which were inconsistent with the well-being of society, and gave a sanction to the pretended art of magic. To everything of this kind the spirit of true philosophy is adverse. It is its just praise, that it not only unfolds to us truths which are suitable to our faculties, but that it shows us within what limits our inquiries should be restricted.

That the Scriptures, in their general tenor, are not unfavourable to the free development of the human mind in the pursuit of truth, is evident from this fact: that wherever they have been permitted to pass unrestrictedly among the common people, the fruits of intellectual liberty have ripened; [and] the arts and sciences have followed in their train. This is the plain testimony of History. Dr. Whately has, in a very remarkable chapter of one of his works, contrasted the genius of the Christian Religion, as it respects the rights of individual minds, with that of Pagan and other systems. He remarks that all religions, but that of Christ, have had two sets of doctrines, the one for the favoured few, the other for the many; whereas the Christian system regards all as alike, acknowledges no caste, espouses no system of reserve.[19] From this remark, which is at once just and profound, we may I think infer, that the Religion within the sphere of whose influence we have the happiness to be placed, is favourable to mental freedom in the abstract; and may assure ourselves, that in the due exercise of our intellectual powers, we are acting in conformity with its spirit and its precepts.

And our confidence in this conclusion will be strengthened if we consider what Truth is. We are not to regard it as the mere creature of the human intellect. The great results of science, and the primal truths of

religion and of morals, have an existence quite independent of our faculties and of our recognition. We are no more the authors of the one class,
than we are of the other. It is given to us to discover Truth,—we are permitted to comprehend it; but its solo origin is in the will or the character
of the Creator; and this is the real connecting link between Science and
Religion. It has seemed to be necessary to state this principle clearly and
fully, because the distinction of our knowledge into Divine and Human,
has prejudiced many minds with the belief that there is a mutual hostility
between the two,—a belief as injurious as it is irrational.

   I cannot entirely dismiss the subject of books without noticing a class
of writers, chiefly on moral and political questions, whose style and manner betray an imitation of the German model. Carlyle and Emerson are
the leaders of this school: and the half idolatrous veneration with which
they are regarded by their disciples, may justify some notice of their peculiarities. We find in their writings many noble thoughts, much force of
language, and an apparent earnestness of spirit; but they are deformed, on
the other hand, by frequent extravagance of language, startling paradoxes,
and positive error and self-contradiction. On the whole, I conceive, that
this school is built upon a false foundation. The investigation of truth is
too solemn, too difficult a thing, to allow of its being associated with a
constant effort to appear striking and original. It is the highest excellence
of the language of true philosophy, to be perspicuous and accurate, and
whenever we find these qualities habitually disregarded, we may suspect
that the writer has some other idol than Truth. Absolutely to question the
sincerity of any class of writers would be unjust or ungenerous, but we
may be allowed to reckon among the causes which tend to pervert the
simple desire for Truth in the mind, the influence of a false model of style
and expression.

   Connected with the subject we are now considering is the general
question of mannerism, not only in literary composition, but also in the
dress, the elocution, [and] the daily habits, of those who are attached to
literary pursuits. There are very few cases indeed, in which this is not to
be regarded as a blemish. Whenever it is the result of imitation, it is an
evident token of weakness; it argues a want of original power. And the
highest ornament of the intellect is truth, so the most genuine grace of
the outward manner is simplicity and unaffectedness,—the visible indications of candour and sincerity within.

I have now but little to add on the education to be derived from books. Many other topics might have been introduced, had it fallen within our scope to enter into particular details. The subjects of history, and of physical and moral science seemed, however, to have the strongest claims upon our attention. Upon the subjects of poetry and fiction it would be unavailing to enter; as our limits do not permit us to give to them that full discussion which their importance in the literature of the present age would seem to require. I will just, however, remark, that I view with some regret, the neglect into which the poets of the last century appear to have fallen. They may not possess the force and vigour which characterize their successors of the present age, or their greater predecessors of the age of Elizabeth and the Stuarts, but in the more equable tenor of their verse, in clearness of thought, and propriety of language, they appear to me in general to excel both. It may be doubted whether those who have almost banished Pope and his contemporaries, have succeeded in supplying that void in popular literature, which they have occasioned.

On the study of languages, the importance of which will sufficiently commend itself to your notice, it is not needful to dwell, nor on the various accomplishments of music and the fine arts, which whenever they meet with a congenial soil, answer a higher purpose than that of mere recreation. If I pass lightly over these subjects, it is not that I undervalue their importance, but that the limits of this address do not permit me adequately to express my sentiments upon them.

To the means of education which exist independently of books, I wish in the next place, only much more briefly, to direct your attention. Setting aside the influence of religion, we may rank as foremost among the agencies of this species of education, the conscientious discharge of the duties of our lawful business and calling. For, the labours of our daily avocation, are not merely the equivalent which we pay in the market of the world for the food and clothing, and other advantages which we enjoy, but also, our sphere of Duty—our field of exercise and trial. Now, every faculty we possess, is strengthened by exercise; and thus when the business of our lives is seen by us in the light of duty, it becomes an important means of confirming the power of that principle and the habit of obedience to its precepts, within us. This appears to me to be a very just view of the design of life, and it is one which invests with a real dignity and importance, the homeliest of our lawful employments.

I have already spoken of the importance of devoting a portion of your leisure to the cultivation of the love of Nature, not only through the medium of books and descriptions; but by actual intercourse with her works. This subject appears to me to be so important, that I may be pardoned for reverting to it. There are two observations on this topic to which I desire more particularly to direct your attention. The first is, that intercourse with the works of Nature tends to correct those false and exaggerated views of life, which a long residence in cities is calculated to engender. We are apt to become too artificial, to conceive of too many things as necessary to our happiness, to forget the simplicity of our real wants in the appliances of luxury, and the grandeur of our real nature amid the external trappings of society. This is not said by way of depreciation, for it is probable that the higher capabilities of the human race, could only have been developed in that state of society, of which the existence of great cities is a characteristic feature. It is merely noticed as an incidental evil. Now the position maintained is, that free intercourse with Nature tends to counteract this evil. Take the most ambitious away from the rivalries of party, and the eager race of competition, or the most artificial from the conventional refinements of society, and place him alone under the broad expanse of heaven, radiant with the beams of the setting sun, or lit up with the silent pomp of night, and for the moment, the objects of his eager striving will appear less in his eyes. There is a silent force in the appeals of Nature to the soul of man, and that force is always exerted on the side of Virtue. It is for this reason that in cultivating the love of Nature, we are at once enlarging the sphere of innocent enjoyments, and planting a powerful auxiliary to all good resolutions within the breast.

The other observation which I wish to make, is, that the influence of external nature is a powerful restorative to the mind that has been overtaxed with labour, or worn with care—that it has a special ministry to soothe the weary, and console the disappointed. How many have turned aside from the tumult of cities, and felt, that they could nowhere recover their lost peace, but among the rural scenes of their infant days,—amid the green pastures, and beside the still waters? This is no morbid sentiment. We feel it to be true in our inmost heart.

Beside these influences, which we may regard as immediate, and independent of the perception of any particular truth or idea, there is a class of moral lessons and similitudes, suggested by the varying aspects of Nature; common to every age, and forming a part of the general education of the human race. Thus the comparison of our life to the leaves of the forest, "Now green in youth, now withering on the ground," is found in both Homer and Isaiah, and recurs perpetually in the stream of poetry from their day to ours.[20] And they who know as little of Isaiah as they do of Homer, who never knew the charm of verse, or the sweet music of song, may have felt the same sentiment rising in their breasts, and touching them for the time with a tender sadness, although they have never given it utterance in words. Who shall say how much of the gentleness that manifests itself even in the most rugged breasts is due to the sweet persuasions of external Nature?

Cultivate then the love of natural beauty. Regard the varied objects which a beneficent hand has thrown around your path, as fraught with instruction, from which you cannot turn aside without incurring the guilt of ingratitude.

The last subject to which I am desirous of directing your attention, as to a means of self-improvement, is that of philanthropic exertion for the good of others. I allude here more particularly to the efforts which you may be able to make for the benefit of those whose social position is inferior to your own. It is my deliberate conviction, founded on long and anxious consideration of the subject, that not only might great positive good be effected by an association of earnest young men, working together under judicious arrangements for this common end, but that its reflected advantages would overpay the toil of effort, and more than indemnify the cost of personal sacrifice. And how wide a field is open before you! It would be unjust to pass over unnoticed, the shining examples of virtue, that are found among the poor and indigent. There are dwellings so consecrated by patience, by self-denial, by filial piety, that it is not in the power of any physical deprivation, to render them otherwise than happy. But sometimes in close contiguity with these, what a deep contrast of guilt and woe! On the darker features of the prospect we would not dwell; and that they are less prominent here than in larger cities we would with gratitude acknowledge; but we cannot shut our eyes to their existence. We cannot put out of sight that improvidence that never looks beyond the present hour; that insensibility that deadens the heart to the claims of duty and affection; or that recklessness, which in the pursuit

of some short-lived gratification, sets all regard for consequences aside.
Evils such as these, although they may present themselves in any class of
society, and under every variety of circumstances, are undoubtedly fostered by that Ignorance to which the condition of poverty is most
exposed; and of which it has been truly said, that it is the night of the
spirit—and a night without moon and without stars. It is to associated
efforts for its removal, and for the raising of the physical condition of its
subjects, that Philanthropy must henceforth direct her regards. And is not
such an object great! Are not such efforts personally elevating and ennobling! Would that some part of the youthful energy of this present assembly might thus expend itself in labours of benevolence! Would that we
could all feel the deep weight and truth of the divine sentiment, that "No
man liveth to himself, and no man dieth to himself!"[21]

The great importance of the subject of Education renders it quite unnecessary that I should offer any apology for making it the subject of a public address.[1] It concerns the common interest that just views of its principles should be diffused throughout society. The parent is concerned because he is responsible for the education of his child; and the child is concerned because the happiness and usefulness and virtue of his future life depend very much upon the way in which his years of preparation are spent; the teacher is concerned because upon the dissemination of just views of the nature of his office depends the estimate and also the reward of his labors. It is in a spirit which has regard to this question as of general interest and of public concernment that I desire to offer a few observations this evening. I have endeavored to divest myself of any private feeling and shall make no special allusions to the practice of my own school, nor draw any illustrations from my own experience as a teacher.

I desire rather to consider myself as advocating views in the wider prevalence of which the interest of all fit and conscientious instructors are involved. As the great extent and manifold bearings of the subject to be considered render it desirable that as little time as possible should be spent in preparatory observations—it will be proper to enter at once upon a statement of the views which I have to maintain.

In the first place then I hold that there are three things to which the attention of the teacher must be chiefly directed—the imparting of knowledge—the producing of a ready command and application of such knowledge by art and practice—and the formation of habits. In the accomplishment of each of these objects, I hold that the order of Nature is to be imitated and obeyed.

Now the order of Nature as manifested both in the discovery and the acquisition of knowledge is an ascending and never a descending order.[2] The child acquaints itself with things before it learns the names of things—it uses its senses before it applies its intellect—it does not begin with the rules of language but with the practice of language—and it is so also in the formation of the moral habitudes. It loves its mother before it knows the name of love—and if well-trained and nurtured, it exercises a certain restraint and self-discipline over the rising of its little passions and troubles before it learns to attach to that exercise a distinct idea or is acquainted with its importance in the entire conduct of life.

This consideration then makes known to us the method in which the elements of knowledge are to be imparted. We must not begin with words. We must as far as possible make the child acquainted with things and it will then feel the want, and understand the use, of words as markers and tokens of the ideas which from things it has acquired. When we cannot procure the things themselves, we must substitute the representations of them, we must appeal to the senses, and make it our first object to give clear ideas—and the words which represent the ideas and the reasons and the rules which exhibit the relations among them, will follow in their due course. Thus in the teaching of geography, our reliance should be, at any rate in all the early stages of the study, upon maps rather than upon books. When a child has before him a pictured representation of the country he is to acquaint himself with, traces its boundaries, follows the courses of its rivers, observes the ramifications of its mountains, sees in visible outline its lakes and bordering seas and every more important natural feature, he gains a clear conception of the subject which he does not readily lose. When the name of that country is repeated to him it is not an unmeaning sound; he attaches to it a distinct class of ideas and associations. With these he connects whatever new information he may receive (for that knowledge is always the most readily received and the most formally retained which connects itself the most aptly with pure knowledge, which finds, if I may so speak, a place in the mind ready prepared to receive it) and thus the stock of his attainments goes on increasing to a harmonious and mutually consistent whole.

Remarks similar to these apply with equal force to the study of history. This should always be associated with the use of maps.[3] The progress of conquest or of colonization should be traced by the pupil himself upon the leaves of his atlas. He must in no instance be satisfied with mere names. Nor is an acquaintance with the mere political outlines of a region sufficient. Its physical details should also be mastered.[4] The progress of historical events in every country of the civilized globe has depended far more on these than is generally believed. Not only have they directly influenced the results of military or industrial enterprise, their indirect action in modifying the characters and tendencies of nations has been no less pregnant with important consequences. I need not point to examples. History is full of them. These considerations however belong to the philosophy of history, a subject which is of the greatest moment in education not so much with a view to the inculcation of any particular system as to the producing of habits of thought and reflection.

This method which I have supposed to be pursued in the teaching of Geography and History may with more or less of adaptation be applied to the teaching of other subjects. We cannot sensibly represent every idea that is to be mastered in the study of grammar, but we can familiarize each by examples. We can teach the young student to compare his own acquired ideas, to make lists of those which agree in some understood particular, and to arrive at the distinctions of speech. I need not pursue the illustration. Further examples will present themselves to everyone who will take the trouble to consider the way in which general conceptions are formed. As from the lives of virtuous men we make clearer and stronger our own conceptions of virtue, and as from the examples of things that are beautiful we heighten our own ideas of beauty, so from the successions of facts and words which are continually passing around him, the mind of the child may be gradually led to understand the principles by which those successions are determined. And this constitutes the stage of transition from sense and observation to science and knowledge.[5]

But to understand is not enough. The most skillful in imparting knowledge sometimes fail to produce scholars because they do not sufficiently connect their teaching with practice. It is but the commencement of the work for the teacher to explain principles—the persevering application of the pupil can alone fix those principles and give readiness in their use.[6] It is my impression that it is in this part of education that modern systems chiefly fail. In all that relates to the communication of knowledge in its full elements—to the removal of difficulties to the smoothing of the paths of learning and if I might so speak, to the strewing of it with flowers, we have seen a great improvement over the methods and the systems of the last generation. But as relates to that patient and continued labor which is requisite on the part of the pupil in order to secure all the higher and more permanent benefits of study, I doubt whether improvement is equally conspicuous. I am not sure that we have not even declined from the standards of former attainment in this respect. This particular view of the question becomes still more important when it is considered with reference to the formation of the habits and the character. Could a system of instruction be formed which should make men learned without any effort on their own part—which should supersede the necessity for thought and exertion—it would nip in the very blossom all the best fruits of a sound education. It is the discipline of patient study that invigorates the mind. It is the strength imparted by contest with difficulties that fits us for conquest over higher difficulties. It is the industrious and conscientious application of our mental powers in a prescribed direction that enables us to subdue our natural impatience of restraint and diligently to apply ourselves to whatever task our future lives may present. I conceive that they who aim to supersede by the perfection of their mechanical teaching all strenuous effort on the part of the pupil, do really, though unconsciously, labor to make superficial scholars and shallow men.

If we examine the order of Nature, we shall perceive that the greatest results do not generally follow the most showy beginnings. The seeds of promise lie hid[den] during the period which is proper to their embryo condition—and develop themselves with the healthy sternness when they do appear. If you force them to a premature expansion, you produce a beautiful but a quickly fading flower. It is very much after this manner in education. That species of training which produces the most brilliant effects in a public oral examination does not lay the best foundation for solid acquirement. It is too often fallacious and injurious—fallacious as a test of real attainment, injurious as to its influence upon the mind of the pupil. But this is a point upon which I do not desire to dwell. I could not have omitted naming it without suppressing my convictions of truth, but I content myself with having named it. And in doing this I think it just to acknowledge that my own principle and practice were formerly different from what they are now.

I hope that the remarks which I have just made are sufficiently clear and explicit, and that the distinct necessity of these three elements of a sound education—the acquisition of clear ideas and just views of first principles, the attainment of readiness and facility in their application, and the discipline of system and order—has been recognized and admitted. I will dwell no longer upon these general considerations, but will limit my remarks to particular subjects, endeavoring to exhibit the principle in the application.

In the teaching of Arithmetic and of its applications in the Arts of Land Surveying, Mensuration, and others, it will readily be seen that that instruction is the most efficient which is the most practical.[7] If in teaching the arithmetic of commerce we could imitate in our school arrangements the transactions of business and make the pupils themselves merchants and traders, we should impart to the study an interest and reality which would not otherwise be attached to it. I do not recommend such a system, because to speculate even with counters, might produce a feeling of cupidity which ought to have no place in the minds of children. But to direct them in the different questions which they may have to consider, to imagine themselves to be in the position of the fictitious parties concerned, and to reason upon their affairs as if they were so, may produce the good result without the attendant evil. In pure arithmetic, the teaching of fractions and decimals may be far more efficiently performed by the aid of sensible representations than without. If we speak of the division of a unit, it may be doubted whether we shall be understood, but if we represent our unit by a line on a blackboard or by a slip of paper and effect visibly or tangibly the operation required, our illustration is at least intelligible. In the teaching of mensuration, we ought not merely to describe the cone, the circle, [and] the pyramid, but to exhibit them in their reality. We ought not only to illustrate every rule by appropriate questions, but to put the line or the rod or the measuring chain in the hands of the pupil and require him to prove his attainments on any fitting object that may present itself. In this way, and perhaps only in this way, his acquaintance with the subject becomes practical, his knowledge apt and meet for any sudden emergency. (These are the principles upon which my own system of instruction in these particular departments of practical knowledge has long been moulded.)[8]

In the teaching of Grammar the same general principles may be kept in view but the form of their application will of course differ. Justly considered, the theory of language depends very much upon the laws of the mental faculties of classification, so that instruction in the science of grammar may be made simultaneous with instruction in the science of reasoning. It is contended by some, that the study of grammar is useless because if the child were accustomed to none but the purest models of language he would learn to speak correctly without rules. I fully admit that this would be the case, and so far as respects the art of using language with propriety, I willingly acknowledge that example is far better than precept, practice then theory. But the almost acquired precision of speech would not lessen the importance of the distinct study of language as a science, and as a mental discipline. Practically it affords almost the only instruction in reasoning that a large portion even of educated countryman receive. I will endeavor to show in what way this object is accomplished. Suppose that a child instead of being required to commit to memory a number of arbitrary definitions were required to write out a catalog of the different objects of his experience, as the sun and stars, trees, plants, furniture and etc., of those of its belief as God, the soul, etc., and lastly of the names of individuals and places—is it not apparent that in doing this he would not only be acquainting himself with the source of the primary distinctions of Grammar, as of Substantives, common and proper, but also learning to think clearly. Again suppose him required to make a list of the qualities which his experience shows to be possessed by those objects, as hot, cold, bright, green and etc., or of the actions which may be conceived in relation to them, as to shine, to move, to grow, do we not perceive that in the forming of these distinctions to which grammarians apply the terms adjective and verb, there are involved not only the means of an increasing acquaintance with the structure of language, but also an exercise of the faculties of comparison and attention—faculties upon the employment of which all the processes of reasoning ultimately depend.

The study of English Grammar should from the very first be associated with the practice of English Composition. The loading of the memory with rules and forms is quite useless without application, and there is no application at once so apposite and so interesting as that which the student provides for himself. Children of the most tender age may gradually be trained to the expression of their thoughts in writing. Descriptions of animals, of the seasons, or of natural scenery may afford subjects for the first exercise of their pen. As they advance to the study of history or geography, new themes will occur, and the language of the books which they read will supply them with appropriate expressions, and give them fluency and facility in the use of language. It is a useful arrangement when such essays alternate with the ordinary school exercises. When each pupil is required to prepare an exercise from a book on one evening, and to write an original essay the following one, a due adjustment of the different elements which contribute to accuracy and to originality of thought is then secured. (I have frequently received as voluntary exercises from children, essays upon descriptive astronomy and similar subjects, which have covered several closely written pages.)

Of the importance of the early and frequent practice of original composition it is scarcely possible to speak in terms too decided. Its highest advantage is that it teaches the pupil to think for himself, but the collateral benefits are almost equally deserving of notice. It is the best exercise in the use of language, it produces correct habits both of orthography and of grammatical expression, and it materially contributes to render accurate the attainments of the people in other departments of study and of practice. One of the best criteria which can be imagined in order to test the clearness and consistency of a person's ideas, is to desire him to express them upon paper. To write clearly we must think clearly—to state in writing our views upon a given subject, it is necessary that those views should be distinctly conceived. And hence there is no method so certain, either for ascertaining the character of other persons' attainments or for forming a just estimate of our own, as the practice of original composition. It contributes much to render this important practice agreeable, and therefore to secure its frequency, if some skill in the mechanical use of the pen is previously acquired. Formerly too much attention was paid to the art of writing, or rather to some ornamental accompaniments of it. In the present day, many whose general views of education are sound and enlightened, have embraced the opposite extreme. The great value of a facile and elegant handwriting is scarcely sufficiently acknowledged. Were those who object to the spending of any considerable time in its acquisition, as to a waste of opportunities for mental improvement, aware of the great service which it may render even in the pursuit of the most abstract studies, their objections would I conceive be diminished. If I may venture to appeal to my own experience, I must acknowledge that if I have met with any success in the prosecution of literature and science, I am bound to attribute it in a great measure to the habit of writing out, early acquired and perseveringly practiced. (I remember, when a youth of seventeen, showing some little essay in mathematics, to a well-trained scientific scholar, and receiving from him the brief but pithy advice, write twice as much as you read. I adopted this advice, and I have never had reason to regret that I did so.) Now the acquisition of a habit of frequent writing, depends much upon the degree of facility with which the mere act is performed. That which we do with ease we usually do with pleasure—and that which is done with pleasure is so far the more likely to be often done. For this reason I should not fear to maintain that the acquisition of a ready and elegant handwriting is an important aid in the prosecution of other and higher branches of education, and that it ought to receive attention as such in addition to his other claims.

Besides the direct advantages which result from the possession of a ready skill in penmanship, there is a species of discipline exercised and its pursuit which is I think worthy of being noticed. Writing, drawing, and other mechanical arts exercise to a certain extent the faculty of imitation and give fixedness to the power of attention, and these are faculties the due training of which is of great importance. A purely intellectual education does not in an equal degree combine these objects. It accomplishes greater ends, but not the same. To this I add that the constitution of the mind is such that it is led to seek [a] variety of pursuits and occupations. When it has been long engaged in a particular way it derives relief from change. The continued exercise of reason or of memory or of imagination becomes at length painful; and mechanical employments are gladly resorted to in preference even to absolute rest. On the other hand the merely imitative arts too closely pursued are felt to be monotonous and uninteresting. The mind recoils from a too unremitting application to them and seeks a stimulus in the exercise of its intellectual power. Both classes of occupation are valuable and each has its place. But their highest and best results are produced by union and admixture. I believe that there are very few studies so remote from each other and so unconnected that they may not in some way be made to contribute to their common furtherance. All Science is founded upon our appreciation of that which is

True. All Art on that of the Beautiful or the useful. All Literature on the one or other according to its character and object. It is improbable that between these elements there should exist any real contradiction. We may not be able to place ourselves in such a point of view as to comprehend the whole in one harmonious prospect, but such a point of view there is. We may be assured of its existence though we cannot reach it. And this consideration should teach us in all we do to keep the common end steadily before us, to regulate all our particular studies or acts, as converging to this result, still adapting ourselves to the bent and capacities of the individual, in the degree of prominence which we give to the particular pursuits.

Upon the study of languages I proceed next to offer a few observations. Of its importance it is not necessary that I should say anything. The remarks which occur to me at the very threshold of the subject is that in ordinary school instruction, the modern languages are usually acquired with far greater rapidity and success than the ancient languages, and this notwithstanding that a far greater amount of care and attention are bestowed upon the latter than upon the former. I suppose that this is a matter of common observation. Although a large portion of the youth of this country receive more or less of instruction in the dead languages, a majority, and I believe a very great majority, of these lose so completely all that they have acquired, as to be unable, a few years after leaving school, to translate a Latin quotation. To what cause shall we attribute this result? Not to the want of knowledge or of industry in the teachers; not in any important degree, to the difference of construction between those languages and our own, although this is undoubtedly an operating cause— but chiefly to the fact that those languages have ceased to be spoken. The practice of our schools, is to begin with the Grammar, to end with the application, we pass from the general formula to the particular instance, not as in the order of Nature's teaching from the particular instance to the general rule. In the schools and colleges of Hungary Latin is still a spoken language, and it is I believe acquired far more successfully than here. A learned friend of mine, a Cambridge scholar, was sometime ago agreeably surprised by the fluent Latinity of an Hungarian sailor. In this country there exist great difficulties for the carrying out of such a design. Many of our words and especially our common words do not admit of immediate translation into Latin. They can only be rendered by a periphrasis, or by some roundabout way of describing our meaning. This circumstance would present an obstacle but not an insuperable obstacle to the adoption of Latin conversation in our classical schools. In schools in which the study of the classics formed but an inferior object, the difficulty would of course be greater. I offer these remarks merely by way of directing attention to the very anomalous position which the study of languages is made to occupy in our existing system of education. Ultimately it may perhaps be seen that in professedly classical schools Latin should be the only medium of conversation, that from others it should be excluded, and that the period of education should be divided between the two according to the taste of the people or the views of his parents, with reference to his future destination. That the prevailing system is not only injurious to the cause of sound learning but in many cases detrimental to the happiness of the rising generation during that period of life in which they are most alive to the impressions of joy or sadness, experience forbids me to doubt. For those who are intended for the learned professions or who possess a decided taste for ancient literature I should be sorry to see the study of Latin and Greek superseded—in the case of others the pursuit of the modern languages and more especially of that copious language of the German, the key to a no less copious literature, will probably with advantage be substituted.

There is another view of the question of the expediency of classical studies which has not received the attention it deserves. It is the moral view. A very large proportion of the extant literature of Greece and Rome and more especially that portion of it which is referred to as the standard of classical elegance is deeply stained with allusions and all too often with more than allusions to the vices of Heathenism. To nearly all the poets and to some portion of the writing of philosophers and the historians this remark applies. I know that there is a spurious delicacy in such matters which creates difficulties where none exist. I know that there is a high and true sense in which to the pure all things are pure. But I greatly doubt whether such considerations are at all applicable in the present case. He who reads for instruction and can bring the results of his reading to the test of settled principles and weigh the actions, the sentiments, [and] the manners of the ancient world in the balance of truth and compare them both as to their measure and their defects with the immutable standard of Truth, may pursue his inquiries without danger. To him there is nothing without its lesson. But that the innocency of youth can be exposed to the contamination of evil without danger I do not believe. In order to judge of this question with impartiality we ought to endeavor to view it from a somewhat different position from that which we at present occupy. Let us imagine that classical studies had never been established in this country— but that our modern literature had reached its present standard of excellence. I am aware that the case is, except in the imagination, an impossible one because our current literature has been more or less formed up on the classical model. Granting however the assumption, should we feel ourselves justified in introducing a new form of literature into the studies of our youth, severe and more accurate it may be as a standard of composition, but tainted with a moral pollution which not only has no place in a Christian literature but cannot even exist simultaneously with it except as a relic of the past. It is a fact significant of the real bearing of this question that the native literature of this age, though in all other respects free beyond that of any former, does not even tolerate by suggestion those things which appear without disguise on the classic page. I do not infer from these considerations that under proper restrictions the study of the ancient classic poets may not lawfully be made part of Education. But in the almost entire absence of such restrictions it appears to me that our most culpable negligence (to use no harsher term) has been manifested. There are I suppose few whose classical reading, if of any extent, does not furnish them with passages in ancient authors which, if right minded, they regret ever to have seen. They feel that no imagined advantages can atone to them for a contact with such ideas. Weighed against considerations of this kind, the glitter of poetic imagery and the refinements of style are as nothing. Among a people deeply impressed with the supremacy of moral ends this would be felt to be true. A Roman poet has told us of the reverence which is due to the minds of the young. We may not forbid them to trace the outlines of a bygone Heathenism, but our permission should not extend to those features the memory of which would have more fitly perished with it, but rather to those struggling examples of greatness and goodness which shine the brighter for the gloom around.

Whatever may be the discipline proper to a riper age, the thoughts of the young should be made familiar only with whatsoever things are lovely, whatsoever things are pure. This leads me to consider more particularly in the last place the questions of moral education—the most

difficult part of the teacher's duty—the highest in dignity and importance. Were I to dilate up on it in terms proportionate to its worth, I
should devote to it more space than to all the other subjects collectively
that I have considered. The limits of this brief essay preclude me from
doing more than touch[ing] upon its prominent features.

I have already spoken of the moral discipline of study and have
described the influence of patience and application in forming the character and the habits. A similar remark may be applied to almost every
other circumstance affecting the condition of the individual in the early
stages of life. Nothing is without[9] significance, nothing in vain. The
example of teachers, the conduct of associates, [and] the character of the
parental home, all produce effects the extent and the duration of which it
is impossible to estimate. Whatever takes place within the little world of a
child's experience, tends to form his character for good or for evil. How
important it is that these many influences should combine for good
alone, [yet] how infinitely more weighty and momentous than any of the
outward distinctions of birth and fortune it is unnecessary to argue. I
assume that in their sober convictions this is felt by all to be true, and I
therefore waive reinforcement of it and pass on to offer some particular
remarks upon the means of attaining the desired end. We are to remember in the first place that children are imitative creatures. What they
observe in the conduct of others they are apt to copy themselves. If they
perceive their teacher to be gentle but firm, patient, self-denying, [and] a
lover of truth and justice, the very observation of these things will do
much to produce in the young minds a love of the same virtues and to
encourage the practice of them in their conduct. It is thus that the humblest laborer in the field of education, unblest with superior talents or
acquirements, may sometimes enjoy the consolation of thinking that he
is contributing to the solid welfare of its own and of a future age. That
which he can effect may be little but it is at least an effect for good. The
testimony of personal character on the side of virtue is perhaps the only
thing in which all classes of instructors can agree, but it is the greatest, the
weightiest of all. It must be remembered that all moral habits are strengthened by exercise. Each act of justice that is done for justice's sake, of kindness that is unaffected and sincere, of conquest over anger or indolence or
any evil propensity, confirms the power of these several virtues in the
breast. It is indeed the principle of moral training. Each step facilitates

that which follows it. Every gain that we accomplish renders the next gain easier. Nor should it be forgotten that the same law applies to the opposite career of deprivation and ruin. It should be remembered that the natural condition of childhood and youth is, according to the designs of their Author, a condition of happiness. This is the beneficent, the divinely appointed order of things, and it cannot be contravened without danger. A happy childhood is an important element in the formation of a virtuous character. Generally when the condition of childhood has been one of suffering, especially if that suffering has been caused by unkindness, the injurious effects of this violation of Nature's order will be felt in afterlife. It is therefore not nearly the dictate of a present benevolent but also of a prospective regard to the future that in all arrangements for the education of the young, their happiness should be consulted. For this reason home education is, under circumstances otherwise the same, preferable to education apart from the kindly influence of the domestic hearth. And here I may be permitted to notice the manifold wisdom which is exhibited in the social constitution of our nature. The parent studies the happiness of his child, he desires its moral advancement and growth in all that is good, but he does not know how much the advancement of the one object promotes the other. He does not perceive that [the] instructive affection which impels him to provide for its innocent gratifications, is also contributing to the production of the kindly affection which shall constitute the blessing of its future days. This much upon the education of the habits and the formation of the character by the influence of circumstances it may here suffice to remark. Of the gradual development of the great principles of morals, of the recognition as distinct fundamental verities, standing out in their own clear light and evidence, undimmed by the mists of error or the perversion of circumstance, it remains to offer a few illustrations.

It is one of the distinguishing characteristics of our human nature that we are enabled to penetrate beyond particular facts and to discern the general principles which underlie them and which give to them their actual form and expression. I have endeavored to illustrate their various instances. Its application in morals is not less clear than in other departments of observation. From the consideration of particular examples of just dealing we are led to the general conception of justice. From individual instances of kindness or of veracity we arrive at the general ideas of benevolence and truth. And so on for the other elements of moral excellence. Some maintain that the office of experience in these cases is to provide the concrete material from which the mind by its own power of abstraction separates that element which is common to every instance. Others on the contrary believe that our experience merely serves as the occasion upon which we become conscious of the fundamental truths of our own moral being. Whichsoever of these views we embrace, the order of the process we are considering remains the same. From the observation of virtue in particular instances we are led to the perception of its essential character and to the recognition of its primal idea. From the practice of it in our own conduct we become impressed with its general obligation and its universal necessity. It is thus seen that the last and highest result of a virtuous education is to make the great principles of justice, of benevolence, of purity and truth and love whatsoever others may be ranked among the primal elements of morals, predominant in the breast. To apprehend them in their generality and distinctiveness—to recognize that sovereignty among the affections and the motives which is their rightful due—and to put in practice the course of conduct which they dictate, with all the warmth of a felt affection and all the facility of a matured habit—this is the result to which our efforts must ever be directed. The reason and the affections—the understanding and the will—must be together engaged. We must approve with our judgment that which is good, we must consecrate to it our affections, [and] we must resolve upon it with all the power of our will. These different elements impart to each other mutual strength and support. Just views and right affections, and an active performance of duty, are not indeed inseparably connected—but they are so allied that the one cannot long prosper without the others. They flourish, [and] they decline together. Lest it should appear to any that I am here offering a mere human in contradistinction to a divine theory of morals, I just add in conclusion that the views which I have stated are in perfect accordance with the teachings of the sacred Scriptures.

For the constitution of the young in rectitude of habits we are taught to train them up in the way wherein they should go. For the duty of cultivating the intellect and the judgment in reference to moral and religious truth, we are directed to prove all things. For the connection of knowledge with practice, we are told to know the will of our Maker and to do it. And for the order in which our duty and our speculative inquiries stand in relation to each other, we are instructed, that he that doeth the will of the Father, the same shall know of the doctrine. There is no error more injurious than that which represents religions and true morals as in any degree opposed to each other. Very different were the sentiments of the venerable Chalmers. In the preface to one of the last publications upon which he was engaged, the edition of the Ethical Lectures of Dr. Thomas Brown, he has given his closing testimony in addition to the many with which his writings abound, to the connection between the great fundamental principles of morals and those of the Christian religion.[10] The primary laws of morality he regarded as the immediate expression of the character of their Author, a view in which it is necessarily implied that they are the foundations of all religious truth—the measure of all the divine dispensations. I shall not further pursue this topic for obvious reasons, but I cannot on the subject of such a vital importance as education suppress my convictions that sound morality and true religion are connected by indissoluble bonds.

May this truth more prevail and may it be felt that religion in the persons of its advocates is then most powerful both for defense against the assaults of error and for aggression upon the kingdom of darkness when it is armed with purity of intention and adorned with innocency of life.

There are few employments of life in which it is not sometimes advantageous to pause for a short time, and reflect upon the nature of the end proposed.[1] The pursuit of knowledge is one of those in which this occasional relinquishment of the field of action for that of meditation, is especially needed. We enter upon it at a period when the feelings are quick and ardent, when the desire of distinction is strong, and when many amiable feelings of our nature, the personal affection which we owe to our dearest friends, and the reverence which seems almost to be due to their very opinions, impel us to engage with eagerness in a task which the all but universal consent of mankind has pronounced to be useful and honourable. There is not one of these motives of which I would desire to weaken the force. But valuable as they are as incentives to exertion, they obviously constitute an incomplete ground for any systematic devotion of our maturer powers. The claims of the pursuit of science, like all other claims with which we are concerned, must ultimately rest upon some intrinsic excellency or special suitableness of the object. Qualities such as these can alone give to it an enduring title to our regard. I design, upon the present occasion, to consider the claims of science in the light of the principle just stated. More especially, I wish to direct attention to the ground of those claims, in the immediate or implied relations of science to human nature; in its relations, namely, as an answer to some of the distinctive wants of the human mind, an exercise to its faculties, a discipline of the character and habits, [and] an instrument of conquest and dominion over the powers of surrounding Nature. In the present divided state of public sentiment, particularly in this country and with reference to this institution, there seems to be need of such a discussion; need also that it should not shrink from occupying the whole field of the inquiry. To ourselves at least it cannot but be useful to endeavour to form an intelligent conception of what is really implied in the pursuit of science, of the spirit which that pursuit demands, and of the ends to which it points. It is proper to state in the outset, that under the term Science, I include all general truths, discoverable by the human understanding, whether they are physical truths relating to the material universe, or moral truths relating to the constitution of our own nature, or truths of any other kind. And the order which I design to pursue is the following:

First, I shall consider the origin of scientific knowledge as respects both its internal and its external sources, and shall briefly examine the nature of its conclusions.

Secondly, I shall, from the previous inquiry, endeavour to draw a just conception of the relations of science to the constitution and design of our own minds, of the benefits which we owe to it, and of the corresponding claims which it possesses upon our regard.

I remark in the first place that all scientific truths are founded upon the observation of facts, that experience though not the only element, is yet an essential element of their existence.[2] The truths of the natural sciences, as of astronomy, or optics, or electricity, are made known to us by the observation of natural phenomena, and by reflection upon the results of that observation. Nor can our knowledge of them be derived from any other origin. Thus every science is, as to its actual progress, a gradually increasing system of knowledge which, beginning with experience, advances ever onward through successive stages towards that perfection which no science has yet reached, which none perhaps ever will reach, but of which the idea becomes clearer and brighter, with every approach that we make. And although in this gradual progress of a science, the necessity for continued observation may become less and less urgent, although in some instances it may even altogether disappear, yet, in every case, it must have supplied the first point of departure. This doctrine which is now so fully acknowledged that to dwell further upon it would be superfluous, did not always meet with acceptance. There was a time in which the indispensable necessity of a foundation in experience for all our knowledge of Nature, was not recognized. Among the ancients it was very imperfectly understood; during the long reign of scholasticism it was all but entirely ignored. And obvious as the principle in question appears to us now to be, it has won its way to general acceptance only through difficulty and opposition. It formed, indeed, with its more important consequences, the chief result of that great review of the sources and the methods of human knowledge, which we owe to the illustrious Bacon.[3] But if science begins with experiment and observation, it does not end with them. All the knowledge which the senses have ever communicated to man, has been a mere collection of facts; and were there in the human mind no powers beyond those to which the senses make their direct appeal, that knowledge would never have advanced to any higher condition than that of facts. With such a state of information, however, the mind does not rest satisfied. It feels the pressure of impulses, it is conscious of the existence of powers and faculties which urge it to reduce the scattered details of its knowledge into form and order. It begins to compare and to classify, and to arrange. It examines in what respects different facts agree, and in what respects they differ; and it inquires how far those differences and agreements are constant; how far they are the results of circumstance or accident. Thus, from the contemplation of facts, the mind rises to the perception of their relations. While in the former stage it is little more than a passive recipient of the impressions of the external world, in the latter it exercises an unborrowed activity. The faculties of judgment, of abstraction, of comparison, [and] of reason, are an agency of strength and power from within, which it brings to bear upon the lifeless elements before it, shaping them into order, and extracting from them their hidden meaning and significance.

Thus, to take one department of human knowledge. It is not enough to have observed the courses of the heavens, the sun and moon, those greater and lesser lights, or that silent and countless multitude of stars, which, as soon as the light of day is withdrawn, unfold before us the true amplitude and grandeur of the material creation. Neither would it be enough, if every phase of apparent change which has swept across the heavens from the beginning of time until now, were recorded for our information. We desire to understand the nature of the phenomena which we survey. We would know how far appearances correspond to realities. We would ascertain the law of that "mystic dance"; we would unveil the secret mechanism of causes which produces the order that is seen, and makes that order perpetual.[4] Impatient alike of unconnected and of causeless phenomena, we would reduce all that we behold around us into subjection to our own understandings. And the frame of nature is so constituted as to permit us in a great measure to accomplish that which we desire. If we are conscious of desires and impulses which cannot rest in the possession of particular and solitary facts, but find only in the contemplation of general truths, of constant and predominating laws, their corresponding goal and end; there is in the constitution of nature that which may answer those desires and satisfy those impulses. If we are sensible of the existence of faculties and powers whose province it is to detect order amid apparent diversity, to discover the indications of cause amid the seeming results of accident, those faculties do not exist in vain. The mind of man is placed amid a scene which can afford to all its powers their appropriate exercise. There is thus a correspondency between the powers of the human understanding and the outward scenes and circumstances which press upon its regard. In this agreement alone is Science made possible to us. The native powers of the mind, cast abroad amid a world of mere chance and disorder, could never have realised the conception of law. On the other hand, the fairest scenes of order, and the most unbroken sequences of causation, would have unfolded themselves in vain before a mind unpossessed of those higher faculties which are necessary to their apprehension. The actual circumstances of our position afford us at once the fairest field of exertion, and the surest guarantee of a success proportioned to the diligence of our labours.

I have dwelt upon this view of the nature of science, because I think it important to our present inquiry that its twofold origin should be fully recognised. If, before the time of Bacon, the external sources of human knowledge were too little regarded, we may, in the strong reaction of a subsequent age against this form of error, discern, perhaps, too much of the contrary tendency. Science, in its actual development, may indeed be compared to some stately temple, whose materials have been brought together from many distant regions, some from the forest, and some from the mine and quarry, but in whose fair proportions and goodly order, we read the traces of the designing mind.[5] To all just theory, experience and observation are indeed the necessary prerequisites, but it is the intellect of man operating by laws and processes of its own, which executes the scheme. To those laws and to those processes there belongs an interest quite independent of the results of material science to which they give birth; and for this reason they deserve a distinct attention. But it is only when the two studies, the material and the mental, are associated together, that the true relations of science to human nature are recognised. Then it is that in studying the laws of external nature, a light is shed upon our own, which may seem to us even of more value than the source from which it is derived.

We have examined the sources from which scientific knowledge is derived. Let us inquire, in the next place, what are the most general conclusions to which it conducts us, with reference to the constitution of the universe.

Science exhibits to us the material or physical universe as a system of being, subject, in all respects, to the dominion of fixed and invariable laws. In that system, to the utmost extent to which either the observations of sense, or the deductions of reason, permit us to judge, chance and accident have no place. The condition of its existence is a rigid, unchangeable, necessity. All its successions are uniform; from its settled order, no deviation is either actual or possible. The courses of the stars are justly said to be appointed; "the sun knoweth his going down."[6] Yet in the conception of this rigorous and dominant necessity, repellent as it would be to our own natures, we see nothing to shock or to offend when viewed in connexion with the idea of matter. If the courses of nature are settled, they are settled in a consistent harmony. If her laws are so fixed that they cannot be broken, they are so fixed in themselves, so fixed in relation to the wants of sentient and intelligent beings, that regularity and beauty are the most conspicuous features of the world over which they preside.

Thus in the most ancient and most perfect of all the physical sciences, Astronomy, we contemplate not a particular orb or system, but a universe of worlds preserved in their mutual order and relation, through the agency of a single prevailing law. We trace the operation of that law in the most diverse consequences. We see it moulding the drops of rain, guiding the stone thrown from the hand in its course, and regulating the swing of the pendulum. We trace it in a larger circle of operations, renewing the waters of the ocean by the healthful play of the tides, deflecting the moon in its orbit, moulding the forms and determining the motions of planetary worlds, larger, and it may be, fairer, than our own.[7] We are taught by the conclusions of analysis, that this law of gravitation has not merely a governing, but a preserving, agency; that it not only determines the motions of the whole system, but so determines them, as to provide for their stability and perpetuity. And beyond the confines of this system of ours, beyond the reach of the unassisted eye or thought of man, science still reads the indications of the same power. The faint lustre of spiral nebulae, and the calculated orbits of double stars, tell us of mightier revolutions, accomplished in obedience to the self-same guiding law.

Or to take a more special illustration of the necessity which governs external nature. A comet suddenly makes its appearance in the heavens. Whence it has come we know not, but we are acquainted with the general laws by which its motions we directed, and know the particular influences to which in this our region of space it must be subject.[8] Three, or, at the most, five exact observations of its position, enable us to apply our knowledge to the determination of its actual motion; or could we by a single observation ascertain its exact place and direction, and velocity, at a given moment, the same end would be accomplished. The future path of the erratic stranger is then marked out, among the constellations. And the course thus assigned to it, it is actually observed to follow, until it again becomes invisible by its remoteness. Here we behold the dominion of necessity. Law is obeyed without choice or alternative. There is no deviation, no shortcoming, no excess. In the more recent and less perfect of the physical sciences, we have similar intimations of the nature and character of the material system. We have no reason to think that the law of definite proportions in the science of chemistry, or the laws of the connected agencies of light and heat, of magnetism and electricity, so far as they are actually known by us, are of any less universal character, than the law of gravitation. Upon each and all of these, once that they are determined, we depend with the conviction of perfect certainty. It has been thought by some that this reliance on the constancy of nature, is an instinctive feeling of the human breast, an original impulse of our nature. But be this as it may, the feeling is one, which if accompanied by a proper estimate of circumstances, is never misplaced. We may be deceived by external appearances, but this source of error apart, the uniformity and universality of the laws of nature is, so far as the range of its just application extends, the most solid foundation of human certainty to which we have yet attained.

We have seen how all the generalisations of science point to the one conclusion, that material nature through all her parts is subject to an inflexible necessity, a necessity which seems to inhere in the very idea of matter, and to be inseparable from all the conditions of its existence. Here then the question suggests itself to us: does the dominion of science terminate with the world of matter, or is there held out to us the promise of something like exact acquaintance, however less in extent, with the interior and nobler province of the mind? The inquiry is twofold, and we may consider it as involving the following questions:

First, whether there exist, with reference to our mental faculties, such general laws as are necessary to constitute a science; for we have seen that it is essentially in the recognition of general laws, not of particular facts, that science consists.

Secondly, supposing that such general laws are discoverable, what is the nature of the relation which the mind sustains towards them? Is it, like that of external nature, a relation of necessary obedience, or is it a relation of some distinct kind having no example and no parallel in the material system?[9] These I conceive to be questions of a perfectly definite character, and it seems to me that they admit of an equally definite answer. First, we are to inquire if the mind is a proper object of science.

That in some sense the moral and the intellectual constitution of man are proper objects of scientific inquiry, must be conceded by all who recognise the existence either of general truths in morals, the knowledge of which may be drawn from our own consciousness, or of any fixed principles in the right operations of human reason. Neither of these can be derived from a merely external source. How varied soever the materials which are brought before the mind, there exist within, principles of thought and reason, which are of common application to them all, and are borrowed from none. There are also certain other principles which are of a more special character, yet, equally with the former, have their seat in the mind. In these principles together are involved the laws of our intellectual nature, even as in the final generalisations of physical science, we discern the laws of the material universe. If it is asked whether out of these common principles of the reason we are able to deduce the actual expressions of its fundamental laws, I reply that this is possible, and that the results constitute the true basis of mathematics. I speak here not of the mathematics of number and quantity alone, but of mathematics in its larger, and I believe, truer sense, as universal reasoning expressed in symbolical forms, and conducted by laws, which have their ultimate abode in the human mind. That such a science exists is simply a fact, and while it has one development in the particular science of number and quantity, it has another in a perfect logic. Now in this view of the laws of our intellectual nature are seen proofs of its relation to science, not less convincing than any which are written upon the physical universe. Similar evidence, though of a less formal kind, is presented in the survey of our moral constitution. Though we are conscious that we often do that which our calmer judgment condemns, not as inexpedient, but wrong; in the very fact of this condemnation we read the existence of some internal rule of right, which indeed we have power to disobey, but which we cannot ignore. To this secret testimony of the heart must be added not only the consenting force of the positive deductions of moral science which are based upon other grounds, but also the full weight of that confirmatory analogy which is drawn from the proved existence of law in our intellectual constitution. The study of Ethics thus becomes an essential part of the study of human nature. We conclude that the mind both in its intellectual and in its moral character is a proper object of science.

Secondly, we are to inquire what relation the mind sustains, to the scientific laws of its constitution.

As it is the office of the laws of reasoning to determine what is correct in the processes of thought, and of the laws of morals to determine what is right in sentiment and conduct, it may safely be inferred that whatever other relations the mind may sustain, it is constituted in some definite relation to those elements which we designate by the terms, Right and True. But in the very nature of these terms it is implied that the relation in question cannot be one of necessary or constrained obedience. Were there no liberty of error, there would be no sense of the peculiar claims and character of truth. There are then rigorous, that is, scientific laws of thought and reason, which are not necessarily obeyed. There are also, however apt to be obscured amid the importunate strivings of interest and passion, eternal rules of right, expressions of the moral character and purpose of their great Author and source. And neither do those exercise upon us any force of actual constraint. But they possess a character and a greatness of their own. They stand before us invested with attributes of reality, and of rightful supremacy, before which every opposing power seems but as a shadow or a usurpation. In these facts are presented to us the distinguishing features of our own higher nature. On its ethical side is freedom, associated with the sense of duty; on the intellectual is freedom, conjoined with the perception of the rightful demands of truth. Let the term, Freedom, be objected to, the fact, under whatever name, remains the same. The optimist may indeed inquire whether a condition of existence liable to error and irregularity, is equally perfect with one from which every such possibility is excluded. But the true idea of human progression lessens, if it does not solve the difficulty. A state of being, whose just action is maintained, and advanced by conscious effort, is felt to be better in itself than all the unintelligent obedience of nature.

I shall not here pause to dwell upon the social and economical sciences, which regard men, not as individuals, but as members of a community, and sharers of a public interest, and which are based upon the consideration of prevailing motives, rather than the requirements of an ideal standard of conduct.[10] As men cannot be divested of their individuality, such sciences do not profess to attain the formal strictness of those which have been already considered. They, however, afford us valuable information as to the general tendencies of society and of institutions; and thus constitute a very important branch of knowledge. It is remarkable with what uniformity those causes operate upon large collections of men, which in the individual seem to merge and be lost amid a variety of conflicting influences. I pass over in like manner some other departments of knowledge, which, depending chiefly upon classification, may be regarded as the precursors of science, rather than science. Let us then revert to what has been said, and endeavour to recapitulate, in a few words, the conclusions which have been arrived at.

Science, then, we may regard as the joint result of the teachings of experience, and the desires and faculties of the human mind. Its inlets are the senses; its form and character are the result of comparison, of reflection, of reason, and of whatever powers we possess, whereby to perceive relations, and trace through its successive links the chain of cause and effect. The order of its progress is from particular facts to collective statements, and so on to universal laws. In Nature it exhibits to us a system of law enforcing obedience, in the Mind a system of law claiming obedience. Over the one presides Necessity; over the other, the unforced obligations of Reason and the Moral Law. Such I conceive to be the true conception of Science. It is a conception in which different elements are involved, partly appertaining to the pure and abstract nature of the object, partly to its more special relation to human conditions. Let us endeavour, from the careful review of these elements, to deduce a reply to the further inquiries,—What are the benefits which science confers? What are the claims it possesses upon our regard?

A narrow estimate of human objects is not likely to be a just one, and even in the sober view of reason, things are valuable upon very different grounds; some things for their own sake, some as means toward the attainment of an ulterior good. It might be well if the actual pursuits of mankind were more often regulated by some deliberate judgment of this nature. Custom, however, and the opinion of others too often prescribe to men what ends they shall pursue, and to what extent they shall follow them. And were it not that in such cases the pursuit often yields that enjoyment which the object sought either does not or cannot produce, we might be tempted to think that the restless strivings of humanity are even more vain than poet and inspired sage have pronounced them to be. Yet, notwithstanding that the aims of mankind are often misdirected, and their bearing upon private happiness yet oftener misunderstood, it is not to be questioned that there do exist ends which are worthy in themselves; worthy of all the expenditure of toil and time which their acquisition demands. To this regard they may be entitled, either as meeting some positive want of our nature, or as tending to some improvement of faculty or character, some essential convenience of life, or other acknowledged good. Now we have seen what is the general conception of Science, as presented in the results of previous discussion. Let us, then, consider some of the particulars involved in that conception, with reference to the question of utility, which is more immediately before us.

We have found it to be one of the characteristics of scientific knowledge, that it owes its origin in part to the desires and faculties of the human mind, and that it bears to them a certain relation of fitness and correspondency. Upon this fact, its first claim to our notice rests. The constitution of our nature is such, that whenever the pressure of the merely animal wants is removed, other and higher desires occupy their place. These are not necessarily to be regarded as modifications of the selfish principle. There is an appetency for knowledge which is not founded upon any perception of utility. Sometimes self-reliant and alone, it exists in solitary strength; sometimes gathering support from human sympathies,

> quemcunque efferre laborem
> Suadet, et inducit noctes vigilare serenas.[11]

I have already remarked how this desire of knowledge lends an impulse to those intellectual faculties, whose province it is to educe general truths, and how the actual subjection of Nature to law affords the means, on an unlimited scale, of exercising those faculties in the most appropriate manner.

Now in these facts it seems to me to be implied, that the pursuit of knowledge, and especially of that kind of knowledge which consists in the apprehension of general truths, is a designed end of human nature. Else wherefore was that desire of knowledge implanted? Or if the feeling be derived, rather than instinctive, wherefore was our nature so contrived that the desire of knowledge should, at a certain stage of advancement, never fail to present itself? Wherefore, too, those faculties which seem to have no other end than knowledge, and which, deprived of their fitting exercise, wither and decline? Wherefore, lastly, that wondrous constitution of external nature, so abounding in lessons of instruction, suitable to our capacity, addressed to our condition? With instances of mechanical adaptation in the works of the Divine Architect, we are all familiar. But to the reflective mind, there are few adaptations more manifest, there is none more complete, than that which exists between the intellectual faculties of man, and their scenes and occasions of exercise. Shall we not then confess that here also design is manifest? And shall not this manifestation of design serve in some degree as the indication of a sphere of legitimate employment, and, where not interfered with by other obligations, of a duty, the neglect of which cannot be altogether innocent?

There are certain further consequences resulting from this office of science as an exercise to our intellectual faculties, to which it may be proper to refer. It is scarcely needful to remark that every faculty we possess, and the intellectual among others, is strengthened by exercise. With respect, however, to the improvement of the individual mind by the discipline of science, it is to be remarked that it implies something more than a strengthening of faculties. It involves also the power of continued attention and the habit of application, the most difficult and most important of mental acquisitions. That the habit is usually an acquired one, is I think manifest, as well as that it belongs to the character rather than to the intellect. Furthermore, scientific studies, besides their direct influence upon the mental habits, instruct us in the right methods of the investigation of truth. For the discovery of truth is not commonly the result of random effort; it is usually, as we have seen, the reward of systematic labour, setting out from the careful examination of facts, and proceeding by definite steps of inductive and deductive reasoning to the evolution of principles. And in this process we need both the precept and the example afforded to us in those great results of accomplished science, which we owe to the patient labours of ages past. Finally, no small accession of intellectual force is due to the deliverance of the mind from that dark prejudice of chance in the physical, of fate in the moral world, to which Ignorance clings with inveterate grasp. Of all the delusions which have cast their baleful shade upon the path of human advancement, this is the most fatal. In the one of its forms it paralyses exertion; in the other it saps the foundation of trust in that righteous appointment which assigns to our actions, even in the present life, inevitable consequences of good or evil. I would appeal to all who have made any study of human motives, whether these are not true representations. Let us, however, bring them to the test of facts.

Careful inquiries assure us that there is a real connection of cause and effect between an undrained, uncleansed condition of our towns, and the prevalence of fever and a general high mortality.[12] I suppose that there are few conclusions better established than this. Every now and then it receives fearful confirmation, when some epidemic disease, making head against all the resources of medical art, emerges from the dark lane or the noisome alley, and sweeps away the rich and the poor in one indiscriminate destruction. Men are, however, for the most part, so reluctant to admit the reality of that which they do not see with their eyes, that this teaching of science, and this confirmation of experience, are sometimes alike void of effect. They cannot perceive with their bodily senses the connection between impure air and disease, and they refuse to believe in invisible laws; or, if they acknowledge them in words, they do not give them any hearty assent. And so the scene of desolation is renewed from year to year.

Nor much unlike the above is the moral scene. No conclusion seems to rest upon a greater weight of cumulative evidence, than that a course of life governed by a consistent regard to the principles of rectitude, is the most favourable to public and private happiness. To carry this conviction into effect, requires that men should in some instances do that which is contrary to their personal interest as judged by common standards. It demands, therefore, that there should be on the part of the individual some trust in unseen principles, strong enough to resist the ever present importunity of appearances. The aspect of society does not, however, present, as its most common feature, this settled regard to principle and calm committal of the affairs of life to its direction. Now, I am very far from saying that such a course of conduct, where it is found, is solely the product of an enlightened understanding; but that this is an important element of the case, is beyond reasonable doubt. The man who has contemplated the subjection of all outward Nature to fixed laws, cannot, when he turns his gaze upon human society, think that its dispensations of good and evil are left to the strivings of self-interest or the scramble of accident. Still less, if he attend to the monitions of the internal witness, if he survey the ineradicable elements of his own being, the self-conscious Will, the authoritative Moral Perception, can he regard those dispensations as the sport of a blind fate, disposing of human affairs as if men were but the wreck and seaweed of a stormy shore. No! The discipline of true science, in disposing us to a belief in general laws, is favourable to a sound morality. If it exalts the consciousness of human power, it proportionally deepens the sense of human responsibility. If it releases us from those meshes of fatalism which bound the ancient Stoic, it is not that it may clothe us with an Epicurean liberty.

But it is not against the prejudices of ignorance alone that Science records her protest. There are dangers not less real in an over-curious spirit of speculation too much exercised in logical subtleties, too little conversant with realities. And against these dangers also the positive results of science constitute the best preservative. The scepticism of the ancient world left no deportment of human belief unassailed. It took its chief stand upon the conflicting nature of the impressions of the senses, but threw the dark shade of uncertainty over the most settled convictions

of the mind; over men's belief in an external world, over their consciousness of their own existence. But this form of doubt was not destined to
endure. Science, in removing the contradictions of sense, and establishing
the consistent uniformity of natural law, took away the main pillars of its
support. The spirit, however, and the mental habits of which it was the
product, still survive; but not among the votaries of science. For I cannot
but regard it as the same spirit which, with whatever professions of zeal,
and for whatever ends of supposed piety or obedience, strives to subvert
the natural evidences of morals, and of that which is common alike to
morals and to religion,—the existence of a Supreme Intelligent Cause.
There is a scepticism which repudiates all belief; there is also a scepticism
which seeks to escape from itself by a total abnegation of the understanding, and which in the pride of its new-found security, would recklessly
destroy every internal ground of human trust and hope. I wish it to be
understood that I do not seek to identify this spirit with any party, or
even personally regard it as co-extensive with any party, but speak of it
abstractly as a temper and habit of the mind, which is commonly, perhaps, the result of a too partial discipline.[13] Now to this, as to a former
development of the sceptical spirit, Science stands in implied but real
antagonism. And as it before vindicated the possibility of natural knowledge, so it now lends all the weight of its analogies in support of the
trustworthiness of human convictions, and the reality of some deep
foundation of the moral order of things, behind the changeful contradictions of the present scene.

The claims of science with which we have hitherto been occupied, are
founded upon its direct relations to human nature, and it is interesting to
notice further the testimonies and indications bearing upon this view of
the subject, which have been left by antiquity. In any inquiry as to what
human nature is, such testimony is perfectly admissible, since in the
records of the thoughts and feelings of a past world, we read but another
development of those principles which are common to our nature in all
periods and under all circumstances.

The instinctive thirst for knowledge, its disinterested character, its beneficial tendencies, are among the most favoured topics of ancient writers.
Cicero dwelt upon them with a peculiar delight, and he has invested
them with more than the common charm of his eloquence.[14] Plato made
them a chief ground of his speculations concerning the just man and the

well-ordered state.[15] Aristotle gave to them the testimony of one of the most laborious of human lives. Virgil devoted the fairest passage of his best poem to the delights of a calm and meditative life, occupied in the quest of truth.[16] Lucretius drew from philosophical speculations the matter of what some have regarded as the noblest production of the Latin muse. Sophocles made Knowledge, in its aspect of power, the theme of incomparably the finest of his choral odes.[17] Aeschylus made Knowledge, in its other aspect of patience and martyrdom, the nobler burden of his Prometheus.[18] And there is ground for the conjecture that such influences were not unfelt by those older poets and seers with whom our own Milton felt the sympathy of a common fate, and desired to share the glory of a common renown. The early dawn, too, of philosophy, not to speak of its subsequent and higher development in the schools of Athens and Alexandria, is full of suggestive indications. Some records, scattered indeed, and dim, and fragmentary, still exist of the successive attempts which were made in Ionia, in the cities of Southern Italy, in Greece, to penetrate the mystery of the Universe, to declare what it is, and whence it came. In those speculations, vague as they are, we discern the irresistible longings of the human mind for some constructive and general scheme of truth, its inability to rest satisfied with the details of a merely empirical knowledge, its desire to escape into some less confined sphere of thought, and, if it might be, to hold "converse with absolute perfection."[19] Nor are the efforts to which such feelings gave birth to be regarded as accidental or unmeaning. They had a prospective significance in relation to the Science that was yet to appear. They were like the prelusive touches of some great master of harmony, which serve to awaken the feeling of expectancy and preparation. I affirm, and upon deliberate examination, that the peculiar order of the development of human thought which preceded the rise and growth of modern science, was not an arbitrary thing, but is in its main features susceptible of explanation. Though for any elucidation of the phenomena of nature, it is utterly worthless, upon the human faculties it throws a light of illustration which can scarcely be valued too highly (a).[20]

Beside the claims of science which are founded upon its immediate relations to human nature, there exist others of an implicit character, which nevertheless give to it far more evident material importance. I speak of its bearing upon the arts of civilised life.

We have seen how science testifies to the fact that the material creation is governed by fixed laws. Upon this truth rests the peculiar value of science as a minister to human wants, and a subjugator of the powers of nature to the will of man. All the operations of art and mechanism, which are but applied science, presuppose this constancy of nature. Because the vapour of water manifests certain constant properties of elasticity and capability of condensation, the steam-engine is possible. Because the laws of magnetic action are fixed, the compass is available for purposes of navigation. Because different species of glass have different dispersive actions upon the coloured rays of light, the achromatic telescope lends its aid to our vision. Because electricity freely traverses metallic wires, and in so doing manifests certain properties of attraction and repulsion, we are able to communicate with the absent by the electric telegraph. In a similar spirit of reliance upon the faithfulness of nature, the husbandman commits his seed to the ground, waiting till the genial influences of sun and shower shall mature it into a harvest. And such is the multiplied industry of man. To this it may be added, that the more that industry is under the control of science, the more does it consist in simply arranging the train of natural circumstances; the inherent and impassive forces of matter ever offering themselves as substitutes for animal toil and animal suffering. To this extension of man's dominion over the inorganic world, there is no visible limit. The properties of matter, both mechanical and chemical, seem to be exhaustless in their variety, knowledge being the key to unlock their uses.

Accordingly, it has been thought by some that the results of science, conjoined with other agencies, open before the human race a career of indefinite progression. They anticipate a period when the physical evils which afflict our present state shall exist no longer, or exist in such measure only as is inseparable from a condition of mortality; when painful toil shall have been replaced by the appliances of mechanism; when the most prolific sources of disease, as crowded cities, undrained swamps, [and] pernicious indulgences, shall have disappeared before a more enlightened study of the conditions of health, and a truer appreciation of the ends of life; when the excessive inequalities of wealth, and the miseries which they entail, shall have yielded to a better moral or social economy; and when the effects of those casualties which prudence cannot avert, as earthquakes, tempests, [and] unfriendly seasons, shall either be reduced to a minimum of amount, or shall be so distributed as to fall with the least oppressive weight upon the community at large. They anticipate that in this happy state of things to come, relieved from the oppressive bondage of physical wants, man shall be at liberty to accomplish, and actually shall accomplish, the higher ends of his being; that while the earth shall shine with more than its pristine beauty, the human family shall not only be clothed with the fair assemblage of the moral virtues, but shall add to them that crown and safeguard of knowledge which has been won from the hard experience of ages of error and suffering.

Speculations of this kind are abused, if they only minister to the sense of human power and pride. They have their use when they instruct us, by the comparison of our actual attainments in the measures of a just and happy life, with that ideal standard to which reason and religion point. Let us ask ourselves why that better condition of things is so far from being realised. The probable conclusion will be, that the impediment is not in any invincible repugnancy in the laws of material nature, nor in any want of power and energy in the human intellect. There seems in the present day to be even a superfluous activity of invention, busying itself to accomplish ends that are not valuable, and ministering to a fantastic vanity. Here, then, we are brought again to that position around which all speculations concerning the true welfare of our species seem to revolve, viz., that it essentially contains a moral element.

But to turn this discussion to some practical issue. Whether that higher state of good shall be realised upon earth or not, they who devote themselves to the pursuits of science will not err, if they keep the prospect of it before them as the scope of their practical efforts. Though contemplation is one end of knowledge, action is another; and if the spirit of science is concentrative in its individual efforts, it is generous and diffusive in its wider aims. I speak here, however, of general tendencies. To make immediate utility the sole guide of scientific research, would defeat the object in view. Let there be a liberal union of the love of truth for its own sake, and the desire to make that truth serviceable to the world, and the chief ends for which knowledge is valuable, will be secured together.

I have now endeavoured to impart to you my own views of the nature and claims of Science. I have in doing this been careful to avoid all exaggeration, believing that the moderation and the exactness which characterise Science should be manifest in its advocacy. What I have thus sought however imperfectly to portray, is some faint image of Truth, partly in her essential lines and features, partly in her immediate aspect and relation to ourselves. And I would now ask you, if after all deductions for the imperfection of the sketch, there is not something in the object that should command our rational esteem; something that may even justify, if I may be allowed the expression, a sober enthusiasm: not that transient blaze of feeling of which, too often, the ashes alone survive, to embitter regret, when the freshness of life's most precious years is irrecoverably gone; but that ardour of quiet and steadfast energy which addresses itself to great ends, knowing their difficulties and patiently subduing them. Such has been the feeling of all who have accomplished any eminent good, whether for their own or for a future age, of all the great masters in art and letters, in science and legislation. Such, in the more humble sphere that has been allotted to ourselves, is the feeling that we should strive to cultivate.

And if, in conclusion, I might say a few words of more special application to this country, and to present circumstances, I would remark, that though to choose or to reject the offered benefits of knowledge is a point within our own election, it is an error to suppose that the conduct of any individual or of any society in this matter can affect its final issues in the world. There may be periods in which the prospects of science, and with them those of human improvement, are sufficiently discouraging. The strong tide of party may set against it. Detraction may assail its friends, misrepresentation sully and distort its beneficent aims. Nevertheless, it is not given to such principles and to such means to accomplish any permanent triumph. Calumny shall not prevail forever. Violence and injustice shall not always usurp the place of reason. There shall be a time when men shall be judged according to their spirit and their deeds. And then shall Truth assert her rightful claims. Science shall vindicate her divine mission in the increase of the sum of human good. Obscured by the mists of prejudice, forgotten amid the strife of parties, she but the more resembles those great luminaries of heaven, which pursue their course undismayed above the rage of tempests, or amid the darkness of eclipse.

NOTE (A)

The constant effort of philosophy in her earlier stages was to establish a basis for a purely deductive system of knowledge. This, which is the final result of united experience and science, was the first aim of speculative thought, antecedent to all true science and to all exact experience. Destitute of these aids, there seems to have been but one mode in which the human mind could proceed in its quest of philosophy, viz., by projecting its own laws and conditions upon the universe, and viewing them as external realities. Such appears to me to be the true ground upon which the earlier phases of the Greek philosophy are to be explained.

The prominent idea of the earliest schools, the Ionic, the Eleatic, &c., was that the universe was a unity. They differed in their account of this unity, variously explaining it by water, air, fire, intelligence, &c.; but the existence of some fundamental unity, comprising the whole of phenomena, was, in perhaps all of them, an agreed point. The terms unity and universe, seem to have been almost regarded as convertible. The pantheistic language of Xenophanes, who, "casting up his eyes to the whole expanse of heaven, declared that the One was God," is a type of their most prevalent cast of thought—Aristotle, *Metaphysics*, i. 6.

In a subsequent stage of philosophy—subsequent in the order of thought, and for the most part in that of time also—there was superadded to the above conception of unity as a ground of phenomena, that of a fundamental dualism in Nature. Existence was viewed as derived from the blending or the strife of opposing elements—good and evil, light and darkness, being and non-being, matter and form, &c. To the latest periods of speculation in the ancient world, these modes of thought, of which the Manichean doctrine was but the most eminent and most practical instance, prevailed; and in those modern schemes of philosophy, "falsely so-called," which attempt to deduce the knowledge of Nature, a priori, from some purely metaphysical principle, the same influence is apparent. Now, so wide an agreement, even in what is false, must have some foundation in reality, and ought to be regarded as a misapplication of truth rather than as a fortuitous coincidence of errors. The foundation must be sought for in the ultimate laws of thought, and the positive conclusions of science serve to show its real nature.

All correct reasoning consists of mental processes conducted by laws which are partly dependent upon the nature of the subject of thought. Of that species of reasoning which is exemplified in Algebra, the *subject* is *quantity*, the laws are those of the elementary conceptions of quantity and of its implied operations. Of Logic, the subject, is our conceptions of classes of things, represented by general names; the ultimate *laws* are those of the above conceptions and of the operations connected therewith. Let these two systems of thought be placed side by side, expressed, as they admit of being, in the common symbolical language of mathematics, but each with its own interpretations—each with its own laws; and together with much that is obviously common—so much, indeed, as to have fostered the idea that Algebra is merely an application of Logic, there will be seen to exist real differences and agreements hitherto unnoticed, but not without influence on the course of human thought. The conception of the universe in the one system will occupy the place of that of unity in the other, not through any likeness of nature, as was once supposed, but through subjection to the same formal laws. Moreover, at the root of the logical system, there will be found to exist a law, founded in the nature of the conception of "class," to which the conceptions of quantity, as such, are not subject, and which explains the origin, though it does not furnish the justification of the dualistic tendency above adverted to. I conceive it unnecessary to show, that a law of the mind may produce its effect upon thought and speculation, without its presence being perceived. Whatever, too, may be the weight of authority to the contrary, it is simply a fact that the ultimate laws of Logic—those alone upon which it is possible to construct a science of Logic—are mathematical in their form and expression, although not belonging to the mathematics of quantity.

My apology for introducing in this place observations of a somewhat technical character is, that in discussing the relations of science to human nature, it seems necessary, or at least desirable, to consider the subject in the light of past as well as of present experience; and to this end, the study of the logical or pre-inductive stage of science is important. But there is also a great collateral interest in the inquiry. The truly scientific study of the laws of thought sets in clear view the distinctive elements of our intellectual constitution—its subjection, like external Nature, to mathematical laws—the difference of the kind of subjection manifest in the two cases. It would seem that this is a fundamental difference. If we strive to conceive of our nature, in its most perfect state, the intellect assenting only to what is true, the will choosing only what is good, the consciousness that all this might, without any violation of our actual constitution, be otherwise, would appear to be a necessary adjunct to that conception. The view of this subject maintained in an earlier portion of the lecture, seems to me to be thus in strict accordance with the proved results of science.

Milton, in one of the most beautiful of his sonnets, addressed to his friend Cyriack Skinner, commends the cheerful wisdom of refreshing the over-wrought mind and the anxious heart by social intercourse.[1] "Let Euclid rest," he says,

> Let Euclid rest, and Archimedes pause,
> And what the Swede intends, and what the French.

And the brief holy day thus snatched from the dust and toil of life, from studies, and cares, and political anxieties, he exhorts him to devote to mirth and friendship. "Heaven," he adds,

> Disapproves that care, though wise in show,
> That with superfluous burdens loads the day,
> And when God sends a cheerful hour, refrain.[2]

He does not tell his friend that he is to neglect the duties of a scholar, for that seems to have been the special vocation of the man, and of a patriot. But he reminds him that there is a time for other things than these—a time for those delights which have been annexed to the companionship of our fellow-creatures—delights the capability of feeling which makes man pre-eminently a social being.

The words of our great poet admit of a wider application than was directly intended for them. The labours and cares of life must, perhaps always, engross our chief attention. And there may be times which add to this ordinary weight of care, a special burden of their own. Milton's friend lived in such times, and we, it seems not unlikely, are entering upon a similar period. To the increased pressure upon the means of life are now added anxious solicitudes about our country, the interests of liberty in Europe, "And what the Swede intends, and what the French."

Cast down by such thoughts, we may need to be reminded, that when we have done all that we can do to provide, as members of families, for the interests of those dependent upon us, as citizens, for the honour and welfare of the state, one business of life yet remains, and that is, "to live." I include under this term the cultivation of our faculties and of our being, the delights of human fellowship, the innocent enjoyment of those good things which have been provided for us in the works of nature and of art. The erection in the city, Cork, of this spacious and beautiful room, designed for purposes of public instruction and amusement, is, in my opinion, a noble expression of the feeling which I have endeavoured to describe. And its public opening, last week, by the chief person in this land, with that befitting state and ceremony, and, as I think I may also add, with that knightly grace and courtesy, which reminded us of olden days, may be accepted as an omen, that the institutions which may spring up under its shelter, shall be truly national in their character, and permanent in their influence for good.[3]

It is, I trust, with the same feelings that we have met together this evening. The members of the Society, which I have the honour to represent, sought not, when they issued their invitation to you, for any occasion of display. But they felt that it was a good thing that men should meet together in amity. And they believed that in the present instance that spirit of amity and of mutual satisfaction would be promoted, if they should succeed in collecting together a few of those objects which exemplify the progress of the peaceful arts, or open up to us the life of other times, or are in some way connected with the interests of humanity. In how kind and liberal a spirit their desires have been seconded by all those who have been asked to render their assistance upon the occasion, I need not describe. The evidences of it are before you. But this you will feel with me, that such a spirit of cooperation both deserves our thanks and commands our esteem. It bears the stamp of a genuine benevolence, when they who possess objects upon which they set more than a pecuniary value, and from the contemplation of which they derive pleasure, are willing to lend them in order that others may participate in their enjoyment. And here I cannot but notice that such benevolence meets with the reward which it deserves, but which it does not directly seek. For the

pleasure which we derive from the contemplation of any great, or curious, or beautiful work, is essentially of that kind which is the most remote
from selfishness. All such pleasures are enlarged by sympathy. I think that
this is more eminently true with respect to the productions of art, than
with respect to the great works of nature. And perhaps that scene which
is the most august of all, demands for its realisation the silence and soli-
tude, as well the darkness of the night.[4] With the productions of human
skill or genius we stand, however, more on a level. We can better understand their purpose and more fully realise the idea which they are
intended to convey. Whether, however, these observations are just or no, it
remains, I apprehend, an undoubted truth that the pleasure which an
occasion like the present inspires, rests very much upon the union of the
social feelings with the enjoyment of the presence of those objects which,
as they are the fruit, so they are the symbols, of peace and good will
among men.

The interesting and beautiful articles illustrative of various departments of art, science, and antiquities which have been contributed for
our instruction this evening, will also serve to explain in some degree
the nature and objects of the Cuvierian Society. That society, as its name
indicates, was originally established for the prosecution of the study of
natural history. Its field of inquiry has, however, since then enlarged, and,
at present, no limits are recognised but these which good sense prescribes. It devolves upon me, this evening, to give a brief sketch of our
proceedings during the past session. Our most important contribution
in Natural History was an account by Professor Murphy, of the river-
weed Anacharis Alsinastrum, a growing specimen of which has been
kindly furnished for inspection this evening.[5] A good deal of attention
has been paid to the subject of archaeology. Something of the life and
national character of ancient Ireland, presented to us in its domestic relics, its manuscripts, and its funeral inscriptions, has been brought before
us by one who has earned the right to speak with authority upon these
things—Mr. Windele.[6] Beautiful drawings, and a lucid description of the
remains of Ardfert Cathedral, in the County of Kerry, have been furnished by Mr. Brash, and the fruits of laborious researches into ecclesi-
astical and family records, by Mr. Caulfield, a gentlemen whose ability
and modest merit are deserving of a position in which he could better
serve the cause of archaeology.[7] Our much respected friend Mr. Sainthill

has made us acquainted with some of the rarer branches of Numismatics, more especially, with the Parthian coinage.[8] The records of an antiquity even more remote have been interpreted to us by Professor Harkness, who has conducted us in thought through those primeval forests to which we are indebted for our fields of coal, and has by his careful sections and original observations taught us something of the geology of the wild western coast of Ireland.[9] To Professor Shaw, we are indebted for illustrations of some of the most interesting of the modern implications of science, more especially in that department of electricity which seems to hold out such wondrous promise.[10] To our friend, Mr. F. Jennings, we owe a record of personal observations in the Island of Cuba, which must, I conceive, have been listened to with profound attention, and with a painful interest, by all who heard them.[11] Many other interesting communications have been made by members of the society, to which I am unable to refer in detail. It would not accord with the feelings of the gentlemen to whom we have been indebted for so many evenings of instruction and delight, neither could I undertake the task, of measuring and apportioning the obligations under which they have placed us. I would rather express my conviction, a conviction in which I am sure many will join with me, that although their communications have been always original, and in some instances the fruit of considerable labour and research, they have been brought forward with a total absence of pretence and affectation, and without thought of self. Some of these researches will, I venture to hope, reach the public through other channels, for as a society we publish no transaction, and aim at little more than social intercourse, somewhat refined by those subjects of thought and conversation whose interest is not merely of the present time.

These few details will serve to show to you that the present meeting involves nothing inconsistent with the design and objects of the society with which it originated. Nor is this the first occasion on which that society has sought to bring itself into more immediate connexion with the social spirit and intellectual activity of the City of Cork. Meetings of the same kind have been held in former years, and the opening of the Queen's College was thus celebrated. Among those who took an interest in their promotion, I may mention that good and honorable man Colonel Portlock, now no longer resident among us.[12] The principle upon which such meetings rest, I conceive to be this, viz., that within proper limits, and under proper conditions, intellectual tastes are not only compatible with social enjoyment, but tend to refine and to enlarge that enjoyment; that an interest in the progress of the arts and sciences and in the researches of the antiquary and the scholar is calculated not to destroy but to deepen our interest in humanity. This is a principle which ought not to need any vindication in the present day, nor shall I attempt its formal vindication. But I will, with your permission, make it the basis of a few remarks which have occurred to me in its connexion with the present meeting.

I would begin by asking you what we mean when we speak of the human race, is it merely so many men and women, isolated units of humanity; some dwelling in this quarter of the globe, and some in that; some enjoying their brief tenure of existence under one of the great planetary cycles, and some under another? You may have stood on a summer's day by some placid lake, and observed, as a light breeze swept by, raising its surface into ripples, how, in obedience to a physical law, each wavelet pursues its own course without interfering with, or in any way influencing the others. You may, in particular have noticed how, when reflected back from the shore, they cross and override those which they meet, but still without mutual disturbance, until they are finally lost and no trace of them is left. Now, can this be taken as a just emblem of human life? Are we who are assembled here, and all who in past ages have felt the joys and sorrows of humanity, but mimic billows upon the sea of time which follow in perfect independence their several tracks, and then dying away, leave its surface as if they had never been? I suppose you will agree that this would not be a true picture of our state and condition here. You will be conscious of the existence of bonds by which each age and each country stand connected with all others. You will feel that there is such a thing as *humanity*. I would beg most distinctly to say that I do not use this term in a sense in which it has sometimes been employed of late, and which seems designed to imply that there is nothing higher and greater than the collective race of man. Perhaps it is in the thought that there does exist an Intelligence and Will superior to our own, that the evolution of the destinies of our species is not solely the product either of human waywardness or of human wisdom; perhaps, I say, it is in this thought, that the conception of humanity attains its truest dignity. When therefore I use this term, I would be understood to mean by it the human race, viewed in that mutual connexion and dependence which has been established, as I firmly believe, for the accomplishment of a purpose of the Divine Mind. And having said this, rather with a view to prevent any possible misconception, than because I think such a theme proper to be discussed upon the present occasion,—I remark that one eminent instance of that connexion and dependence to which I have referred is to be seen in the progression of the arts and sciences. Each generation as it passes away bequeaths to its successor not only its material works in stone and marble, in brass and iron, but also the truths which it has won, and the ideas which it has learned to conceive; its art, literature, science and to some extent its spirit and morality. This perpetual transmission of the light of knowledge and civilisation has been compared to those torch races of antiquity in which a lighted brand was transmitted from one runner to another until it reached the final goal. Thus it has been said do generations succeed each other, borrowing and conveying light, receiving the principles of knowledge, testing their truth, enlarging their application, adding to their number, and then transmitting them forward to coming generations—"Et quasi cursores vitai lampada tradunt."[13]

Now, this connexion between intellectual discovery and the progressive history of our race, gives to every stage of the former a deep human interest. Each new revelation, whether of the laws of the physical universe, of the principles of art, or of the great truths of morals and of politics, is a step not only in the progress of knowledge, but also in the history of our species. Could we trace back our intellectual pedigree, if you will permit me to use such an expression, we should find ourselves connected by that noblest of all lines of descent, with every nation and kindred of men that has occupied a place in history, and with many others, of whose names and deeds no record survives. We should see the picture writing, most probably of some forgotten Asiatic tribe, passing through successive stages, analogous to those which are still preserved in the monuments of Egypt, until among the Phoenician people it gave birth to our present system of letters. We should behold the first principles of our science, and much more than the first principles of our literature and philosophy, emerging into light among those isles of Greece, which seem to have been the chosen home of freedom and of genius in the ancient world. To the same source we should trace back whatever is most refined in the art of the sculptor, and no small portion of the science of the architect. To the Romans, above all others, we should find ourselves indebted for the principles of government and law. Theirs was even less the genius of conquest than of empire and rule; and the system of juris-prudence which they have left is still, in the opinion of some, their noblest monument. To the Arabians we owe our numerals, and through this the science of arithmetic. And beside these more distinct portions of the inheritance which has been transmitted to us from ages past, and of which the enumeration is far from being complete, how many customs, thoughts, and opinions, how many silent influences for good or for evil, do we not unconsciously owe! As respects the larger and more definite accessions to which I have referred, it would almost seem as if the law of human progression were this—that to different sections of the one great family of man, different measures of special capacity were assigned, so that each, while fulfilling its own destiny, should also add to the common stock of intellectual wealth.[14] I conceive the Greek art to be an eminent illustration of this principle, though others, scarcely less signal, might be adduced. Thus, it has been doubted whether we could, in the present day, originate that union of wildness and romantic beauty, of grotesqueness and grandeur, which constitutes the predominant character of Gothic architecture.[15] I can well conceive that it was only from a certain order of mind, the ground of whose character was formed amid the pine forests of the north, and whose later stamp was received from the stately but decaying monuments of Imperial Rome, that such a product could have arisen. But, having come into being, it remains, through its works and its conceptions, the parent of solemn thoughts to all succeeding times. There is, I need not remind you, one special task which these later ages seem destined to accomplish, a task of the highest importance, but which it would be a fatal error to regard as an end, and not as a means; it is the extension of man's dominion over the material world. I will not attempt to examine here the various aspects of that much-disputed question, why so subtle and inquisitive a people as the Greeks made no advance in physical science?[16] It has been said that it is because they did not possess a proper method. But the difficulty is thus thrown back and not solved—for the question immediately arises, Why did they not possess a proper method? The principles of that method are so obvious as to be almost axiomatic, and in other departments of speculation they were under-stood and applied by some of the great thinkers of antiquity. I suppose that we must conclude, with an eminent writer on the history of the inductive sciences, that the time for this development was not come, that there were other problems to be solved first, more intimately connected with human freedom and happiness. In confirmation, however, of the fact that the extension of human sway over the material world is an actual, whether or not it is a special business of these times, we have only to consider what is going on around us, and has been exemplified in this room. Take, for instance, that science of chemistry, which imitates, by its processes, the products of organic life, or those improvements in metal-lurgy, and those applications of the wonderful power of electricity to the arts, some of which have been elucidated by my friend and colleague, Dr. Blyth, this evening.[17] Or reflect upon those recent advances in the science of optics which have brought to its present state the achromatic microscope, an instrument which enables us to survey some of the most secret processes of nature, and which, aided by that peculiar modification of light which is produced by the action of certain crystals, permits us in some instances almost to look into the chemical constitution of bodies.[18] Or take that latest birth of science, the art of Photography, which makes its pencil of the solar beam, and delineates, with a faithfulness to which no human limner can approach, not merely the repose of nature, but her aspects of change, the waves breaking on the shore, or the human coun-tenance lit up by some moment, gleam of thought or feeling.[19] These examples have been suggested to me by objects in this room. I forbear to multiply them, or to refer to the mechanical triumphs of the age, lest in the multitude of the illustrations you should forget the principle which they are designed to exemplify.

There exists yet another and not less important view of the nature of those elements which constitute civilisation. It is that the progress of knowledge and the arts not only forms a bond which connects the dif-ferent generations of men together by interests and feelings wider than those which are merely national; it serves also as a progressive manifesta-tion of the nature of man,—it makes us acquainted with the hidden capacities of our being. I remember the profound interest with which I read, some years ago, a treatise by a German writer, written with that fulness of learning which the Germans alone possess, and also with that ripeness of judgment which they do not always display, intended to trace the development among the ancient Greeks of the idea of the chief good of man.[20] The author showed how that idea was associated among the earlier writers as Homer, almost exclusively with the possession of physical qualifications, largeness of stature, strength of limb, swiftness of foot, or with such intellectual endowments as we should now term cleverness, and perhaps cunning. He showed how at a later period it was connected with wealth and longevity, with the glory of ancestry, the exercise of a large and bountiful hospitality, the esteem of men. This is the form which it chiefly assumes in the writings of Pindar. Then he traced the idea through the Gnomic poets, under the form of prudence, self-respect, reverence for law and established religion, until in the conversations of Socrates, it rises to the full measure of the conception of moral good. Now, this picture, though drawn from a source lying a little out of the general line of illustration, which I have adopted, will serve to explain the position I wish to establish. We are not to suppose that there was any moral faculty in Socrates disputing among his friends about the true ends of life, which did not also exist, only in a less developed degree, in the heroes of the Iliad fighting before Troy, and the youth of Greece contending in the Pythian Games. But this is the lesson which I wish to draw, a lesson, too, which upon other grounds has been enforced in this city by one far more competent than I am to do it justice, viz., that it is not in the rude and ignorant, or in the savage and feral state of Man, that we can see what human nature is. Its inferior elements predominate there, and all its nobler and more characteristic qualities remain hidden. It is the slow but combined action of the social state which brings out the germs that would otherwise lie buried beneath a stony and a wintry soil. Science, while it is thus a revelation of the laws of the material universe, is also a manifestation of the intellectual nature of man. So too all those arts which depend upon the perception of proportion, whether it be in forms or in sounds, are at least as dependent upon the existence of certain faculties of our nature, which faculties they make known to us, as upon any relations of external things. What a world of sweet and solemn emotions for instance does not music awaken within us, a world of whose existence we should but for that divine art be wholly unconscious, and of whose possible limits we are still ignorant! It is not in the instrument nor in the pulses of the air, nor in the mechanism of the human ear, that the harmony resides, but in ourselves. In the mysterious depths of the human spirit those faculties have their abode, for whose calling forth all these external movements are but a preparation. And the science of the organ-builder and the skill of the musician consist in this, that they understand, practically at least, some part of that connexion which has been established between mental and material things, by Him who is both the maker of the universe, and the author of our spirits. But I will not pursue this line of argument and illustration further. I trust that whether you agree with it or not you will not dissent from the conclusion which it was designed to support, viz., that the central bond of the arts and sciences must to us at least consist in the idea of humanity, in their connexion on the one hand, with the general progress of our race, [and] on the other, with the development of the nature of the individual. And I think that I do not err in supposing that this consideration gives even to the most mechanical of the processes which have been exhibited to you this evening a human interest, distinct from that which they possess as ministering, according to the language of Bacon, to the relief of man's estate.

I might, if time permitted, take up the remaining branch of the argument, and show that the researches of the antiquary and the scholar possess, when rightly pursued, the same kind of claim to our regard as the labors of the artist and the man of science. Undoubtedly there exists a great deal of trifling curiosity about things of no moment, and many a vain attempt has been made to reconstruct a living form out of those dry bones of antiquity from which the breath of life has fled forever. But all researches into the past are not vain, and I conceive it to be impossible for a man of any cultivation of mind to read without interest the results of such inquiries as have recently been published by the Chevalier Bunsen, into the origin and connexion of languages, and their bearing upon many problems of deep historic moment.[21] In these pursuits, as in all others, but in these more eminently, there is need of a controlling principle. Things are not valuable because they are old and rare; but the interest which gathers about the relics of bygone ages is then only legitimate when it flows from a deeper source—even from the sense of the fellowship of humanity. That is the idea which I wish to convey to your minds.

But I do not forget that this evening, though as closing the session of a scientific society it has called for some discussion of graver matters, was also to be devoted to social intercourse. Whether then I have succeeded in my imperfect attempt, or not, let me express a hope that you have derived pleasure from the many beautiful objects which have been brought before you, and that you have, above all, felt that delight which is seldom unfelt, when we look upon the faces, and hear the voices of friends and neighbours.

NOTES

PREFACE

1. Alexander Wheelock Thayer, Hermann Deiters, Henry Edward Krehbiel, and Hugo Riemann, *Thayer's Life of Beethoven*, ed. Elliott Forbes (Princeton: Princeton University Press, 1991), 57.

2. Edward O. Wilson, "How to Unify Knowledge: Keynote Address," *Annals of the New York Academy of Sciences* 935 (2001): 12; and see his *Consilience: The Unity of Knowledge* (New York: Alfred A. Knopf, Inc., 1998); as well as C. P. Snow, *The Two Cultures and the Scientific Revolution* (Cambridge, UK: Cambridge University Press, 1961).

INTRODUCTION

1. For a typical formulation, see Paul J. Nahin, *The Logician and the Engineer: How George Boole and Claude Shannon Created the Information Age* (Princeton, NJ: Princeton University Press, 2012).

2. Jozef Bocheński, *A History of Formal Logic*, ed. and trans. Ivo Thomas (Notre Dame, IN: University of Notre Dame Press, 1961), 295–297; James Gasser, ed., *A Boole Anthology—Recent and Classical Studies in the Logic of George Boole* (London: Springer, 2000), especially 61–78, 129–166. See also Susanne K. Langer, *An Introduction to Symbolic Logic*, 3rd ed. (Mineola, NY: Dover Publications, 1967), 17–18; W. Kneale, "Boole and the Algebra of Logic," *Notes and Records of the Royal Society* 12 (1956): 53–63; Irving Anellis, review of *The Development of Symbolic Logic*, by Arthur Thomas Shearman, *The Review of Modern Logic* 11 (2007–2008): 87–105.

3. Lewis Carroll, *Symbolic Logic* (London: Macmillan, 1896), republished with manuscript integrations by William Warren Bartley III (New York: C. N. Potter, 1977).

4. Bertrand Russell, "Mathematics and the Metaphysicians" (1901), reprinted in Bertrand Russell, *A Free Man's Worship, and Other Essays* (London: Unwin 1976),

75. For the following aspects, see also Carl B. Boyer, *A History of Mathematics* (New York: John Wiley, 1968), 632–636.

5. See chapter 6.

6. Desmond MacHale, *The Life and Work of George Boole: A Prelude to the Digital Age*, reprint of the 1985 edition with a new preface by Ian Stuart (Cork: Cork University Press, 2014), chaps. 1–2.

7. For this point in Boole's life, see ibid.; Kenneth C. Dewar, *Charles Clarke, Pen and Ink Warrior* (Montreal: McGill-Queen's University Press, 2004), chap. 1.

8. Cited in chapter 6.

9. Denis Lawton and Peter Gordon, *A History of Western Educational Ideas* (London: Woburn Press, 2002), 97–100.

10. For a fundamental work, see Sarah Tarlow, *The Archaeology of Improvement in Britain, 1750–1850* (Cambridge, UK: Cambridge University Press, 2007), 124–162. See also Keith Tribe, "Political Economy and the Science of Economics in Victorian Britain," in *The Organisation of Knowledge in Victorian Britain*, ed. Martin Daunton (Oxford: Oxford University Press, 2005), 115–138.

11. John Stuart Mill, "The Spirit of the Age" (1831), in *The Collected Works of John Stuart Mill*, ed. Ann P. Robson and John M. Robson (London: Routledge and Kegan Paul, 1986), 22:230.

12. Stefan Collini, Donald Winch, and John Burrow, *That Noble Science of Politics: A Study in Nineteenth-Century Intellectual History* (Cambridge, UK: Cambridge University Press, 1983), especially the prologue. See also Christopher Fox, Roy Porter, and Robert Wokler, eds., *Inventing Human Science: Eighteenth-Century Domains* (Berkeley: University of California Press, 1995), chap. 8; Deborah A Redman, *The Rise of Political Economy as a Science: Methodology and the Classical Economists* (Cambridge, MA: MIT Press, 1997), chaps. 1–3.

13. Jack Morrell and Arnold Thackray, *Gentlemen of Science: Early Years of the British Association for the Advancement of Science* (Oxford: Clarendon Press, 1981), 249. The words were Sedgwick's.

14. T. H. Lister (anonymous), *Edinburgh Review* 68 (October 1838): 75, 77. In general, and with a relevant bibliography, see Valerie Purton, *Dickens and the Sentimental Tradition: Fielding, Richardson, Sterne, Goldsmith, Sheridan, Lamb* (London: Anthem Press, 2012), 1–18.

15. For population figures, see James Bell, *A New and Comprehensive Gazetteer of England and Wales* (London, 1835), 3:65; Anonymous, *Accounts and Papers of the House of Commons* (London, 1851), 13:6.

16. Friedrich Engels, *The Condition of the Working Class in England* (London: P enguin Classics, 1987), 29, 268. Compare Eric J. Hobsbawm and George Rudé, *Captain Swing* (London: Lawrence and Wishart, 1969), 152–172. On conditions

in Lincolnshire, see T. L. Richardson, "The Agricultural Labourers' Standard of Living in Lincolnshire, 1790–1840: Social Protest and Public Order," *Agricultural History Review* 41, pt. 1 (1993): 1–20. Concerning impoverishment, see K. D. M. Snell, *Annals of the Laboring Poor: Social Change and Agrarian England 1600–1900* (Cambridge, UK: Cambridge University Press, 1985), 15–66.

17. Georg Wilhelm Friedrich Hegel, "Philosophy of Right," trans. T. M. Knox (Oxford: Clarendon Press 1952), para. 236. The original phrase actually appears in Georg Wilhelm Friedrich Hegel's *Werke*, vol. 8: *Grundlinien der Philosophie des Rechts: oder, Naturrecht und Staatswissenschaft im Grundrisse*, ed. Eduard Gans (Berlin: Duncker und Humblot, 1833). In addition, see Karl Polanyi, *The Great Transformation: The Origins of Our Time* (1944), recently reissued with a new subtitle, foreword by Joseph E. Stiglitz, and intro. by Fred Block (Boston: Beacon Press, 2001). For a deeper contextualization as well as a useful bibliography, see Sandra Halperin, "Dynamics of Conflict and System Change: The Great Transformation Revisited," *European Journal of International Relations* 10, no. 2 (2004): 263–306.

18. See, for example, Thomas Babington Macaulay, *The Complete Writings of Thomas Babington Macaulay in Ten Volumes*, vol. 1, *Miscellanies* (New York: G. P. Putnam's Sons, 1901), 1–19. See also John Clive, *Macaulay—the Shaping of the Historian* (New York: Vintage Books, 1975), 204.

19. Pauline Gregg, *Modern Britain: A Social and Economic History since 1760* (New York: Pegasus, 1966), chap. 8.

20. On the Poor Law and the relative historiography, see George R. Boyer, *An Economic History of the English Poor Law, 1750–1850* (Cambridge, UK: Cambridge University Press, 2006), especially chap. 2. On the political backdrop in general, see Boyd Hilton, *A Mad, Bad, and Dangerous People? England 1783–1846* (Oxford: Clarendon Press, 2006), chap. 3; Jonathan Parry, *The Rise and Fall of Liberal Government in Victorian Britain* (New Haven, CT: Yale University Press, 1993), chaps. 2–3.

21. Gregg, *Modern Britain*, 208; John Saville, *1848: The British State and the Chartist Movement* (Cambridge, UK: Cambridge University Press, 1987), 1–50. In addition, see David Goodway, *London Chartism 1838–1848* (Cambridge, UK: Cambridge University Press, 2002), 12–23.

22. In general, see Neil J. Smelser, *Social Paralysis and Social Change: British Working-Class Education in the Nineteenth Century* (Berkeley: University of California Press, 1991), chap. 8.

23. Gary S. Cross, *A Quest for Time: The Reduction of Work in Britain and France, 1840–1940* (Berkeley: University of California Press, 1989) 32–34. For an informative historiographical essay, see John Host, "Narratives of the Past or Histories of the Present?," in *Victorian Labour History: Experience, Identity, and the Politics of Representation* (Abingdon, UK: Routledge, 2002), 8–59.

24. For an especially illuminating work on this issue, see M. J. D. Roberts, *Making English Morals: Voluntary Association and Moral Reform in England, 1787–1886* (Cambridge, UK: Cambridge University Press, 2004), especially chap. 3.

25. Quoted in Steven Shapin and Barry Barnes, "Science, Nature, and Control: Interpreting Mechanics' Institutes," *Social Studies of Science* 7, no. 1 (1977): 36. In addition, see Steven Shapin, "The Pottery Philosophical Society, 1819–1835: An Examination of the Cultural Uses of Provincial Science," *Science Studies* 2, no. 4 (1972): 311–336.

26. Frederic Hill, *National Education; Its Present State and Prospects* (London: C. Knight, 1836), 2:200.

27. *The Mechanics' Magazine* 21 (1834): 153.

28. *Lincolnshire Chronicle*, Friday, February 16, 1844.

29. *Lincolnshire Chronicle*, Friday, February 12, 1847.

30. The Mechanics' Institutes have not been studied in depth, but especially concerning Avon, Somerset, Gloucestershire, and Shropshire, see Colin Turner, "Politics in Mechanics' Institutes, 1820–1830: A Study in Conflict" (PhD diss., Leicester University, April 1980).

31. Martyn Walker, *The Development of the Mechanics' Institute Movement in Britain and Beyond: Supporting Further Education for the Adult Working Classes* (London: Routledge, 2016), 1–12.

32. Jürgen Habermas, *The Structural Transformation of the Public Sphere: An Inquiry into a Category of Bourgeois Society*, trans. Thomas Burger with Frederick Lawrence (1962; repr., Cambridge, MA: MIT Press, 1989). Concerning this work, see Craig Calhoun, ed., *Habermas and the Public Sphere* (Cambridge, MA: MIT Press, 1992), especially chaps. 12 and 13; also, Rodney Benson, "Shaping the Public Sphere: Habermas and Beyond," *American Sociologist* 40 (2009): 175–197. For a modified view incorporating multiple publics, see Christina Parolin, *Radical Spaces: Venues of Popular Politics in London, 1790–1845* (Canberra: ANU Press, 2010), 8–11.

33. For the connection between faith and science in Boole, see Daniel J. Cohen, *Equations from God: Pure Mathematics and Victorian Faith* (Baltimore: Johns Hopkins University Press, 2007), chap. 3.

34. See, for instance, Gordon C. Smith, *The Boole–De Morgan Correspondence: 1842–1864* (London: Oxford University Press, 1982).

35. On the *Cambridge Mathematical Journal* papers and Boole's role in the early history of invariant theory, see Victor J. Katz and Karen Hunger Parshall, *Taming the Unknown: A History of Algebra from Antiquity to the Early Twentieth Century* (Princeton, NJ: Princeton University Press, 2014), 458–462.

36. George Boole, *The Mathematical Analysis of Logic: Being an Essay toward a Calculus of Deductive Reasoning* (Cambridge, UK: Macmillan, Barclay, and

Macmillan, 1847), 3–7. For a general orientation toward Boole's work in the field, see Theodore Hailperin, *Boole's Logic and Probability: A Critical Exposition from the Standpoint of Contemporary Algebra, Logic, and Probability Theory* (Amsterdam: Elsevier, 1986), 63–134.

37. Letter to William Thomson, dated August 17, 1846, cited in MacHale, *The Life and Work of George Boole*, chap. 5.

38. Letter to William Thomson, dated December 22, 1846, cited in MacHale, *The Life and Work of George Boole*, chap. 5.

39. "The Late Food Riots in Ireland," *Illustrated London News*, November 7, 1846.

40. For an up-to-date account of the extensive bibliography, see John Crowley, William J. Smyth, and Mike Murphy, eds., *Atlas of the Great Irish Famine* (New York: New York University Press, 2012).

41. Anthony Trollope, *An Autobiography* (London: Blackwell, 1929), 47, 51, 59.

42. Robert Kane, *The Industrial Resources of Ireland* (Dublin: Hodges and Smith, 1844), 394, 398.

43. In this regard, see especially Enda Leaney, "Missionaries of Science: Provincial Lectures in Nineteenth-Century Ireland," *Irish Historical Studies* 34 (2005): 266–288.

44. On this topic, see Geoffrey Taylor, "George Boole, F.R.S., 1815–1864," *Notes and Records of the Royal Society* 12 (1956): 51. On the university in general during this period, see John A. Murphy, *The College: A History of Queen's / University College Cork, 1845–1995* (Cork: Cork University Press, 1995), chaps. 1–3.

45. Rush Rhees, "George Boole as Student and Teacher: By Some of His Friends and Pupils," *Proceedings of the Royal Irish Academy, Section A: Mathematical and Physical Sciences* 57 (1954–1956): 74–78.

46. The Saint-Simon episode is recalled in Friedrich Hayek, *Studies on the Abuse and Decline of Reason: Text and Documents*, ed. Bruce Caldwell (London: Routledge, 2010), ch. 12. In addition, see Rebekah Higgitt, *Recreating Newton: Newtonian Biography and the Making of Nineteenth-Century History of Science* (London: Pickering and Chatto, 2007), 1–18.

47. Concerning these two portions of Newton's productivity, see James E. Force and Richard Henry Popkin, *Essays on the Context, Nature, and Influence of Isaac Newton's Theology* (Dordrecht: Kluwer Academic Publishers, 1990); and Betty Jo Teeter Dobbs, *The Janus Faces of Genius: The Role of Alchemy in Newton's Thought* (Cambridge, UK: Cambridge University Press, 1991).

48. Thomas Hodgskin, *Popular Political Economy: Four Lectures Delivered at the London Mechanics Institution* (London: Charles and William Tait, 1827).

49. On the memorabilia market, see Patricia Fara, *Newton: The Making of Genius* (New York: Columbia University Press, 2004), chap. 8. On the tooth, referring to an incident in 1816, see *Mechanics' Magazine* 8 (1836): 272.

50. Mary Everest, "The Home-Side of a Scientific Mind," *University Magazine* 1 (1878): 335. In addition, see Rhees, "George Boole as Student and Teacher," 74.

51. David Brewster, *The Life of Sir Isaac Newton* (London: J. Murray, 1831), reissued in an augmented edition as *Memoirs of the Life, Writings, and Discoveries of Sir Isaac Newton*, 2 vols. (London: Hamilton, Adams and Co., 1855). Concerning the papers, see Sarah Dry, *The Newton Papers: The Strange and True Odyssey of Isaac Newton's Manuscripts* (Oxford: Oxford University Press, 2014), especially chaps. 1–2.

52. Concerning this aspect, see Laura J. Snyder, *Reforming Philosophy: A Victorian Debate on Science and Society* (Chicago: University of Chicago Press, 2010), chap. 1.

53. See J.-B. Biot, "Life of Newton," in *Lives of Eminent Persons*, trans. H. Elphinstone (London: Baldwin and Cradock, 1833); reprinted in Rob Iliffe, Milo Keynes, and Rebekah Higgitt eds., *Early Biographies of Isaac Newton 1660–1885*, vol. 2: *Nineteenth-Century Biography of Isaac Newton: Private Debate and Public Controversy* (London: Pickering & Chatto, 2006), 1–63.

54. See Thomas Galloway, reviews of three French and English biographies of Newton, *Foreign Quarterly Review* 12 (1833): 1–27. See also Brewster, *Life of Sir Isaac Newton*, vi.

55. For more recent interpretations of the incident, see Milo Keynes, "Balancing Newton's Mind: His Singular Behaviour and His Madness of 1692–93," *Notes and Records of the Royal Society of London* 62, no. 3 (2008): 289–300.

56. See Massimo Mazzotti, "The Two Newtons and Beyond," *British Journal for the History of Science* 40 (2007): 105–111.

57. On contemporary dilemmas, see David B. Resnik, "Openness versus Secrecy in Scientific Research," *Episteme* 2, no. 3 (2006): 135–147. More generally, see W. D. Garvey, *Communication: The Essence of Science* (New York: Pergamon Press, 1979). See also Jan Golinski, *Science as Public Culture: Chemistry and Enlightenment in Britain, 1760–1820* (Cambridge, UK: Cambridge University Press, 1992), 1–10.

58. On the topic, referring mainly to pre-1700, see Pamela O. Long, *Openness, Secrecy, Authorship: Technical Arts and the Culture of Knowledge* (Baltimore: Johns Hopkins University Press, 2003), 244–250.

59. Stillman Drake, "Galileo, Kepler, and Phases of Venus," *Journal for the History of Astronomy* 15, no. 3 (1984): 198–208.

60. For interesting reflections and a bibliography on Newton's practices, see Koen Vermeir, "Openness versus Secrecy?," *British Journal for the History of Science* 45, no. 2 (2012): 165–188; Niccolò Guicciardini, *Isaac Newton on Mathematical Certainty and Method* (Cambridge MA: MIT Press, 2009), 348.

61. Jacob Bryant, *A New System or an Analysis of Ancient Mythology*, 3 vols. (London: T. Payne, 1774–1776). Here I have used the six-volume edition (London: J. Walker, 1807). A Boole lecture on polytheism for the year 1841 is mentioned in George Oliver, *An Account of the Religious Houses, Formerly Situated on the Eastern Side of the River Witham* (London: R. Spencer and C. W. Oliver, 1846), x.

62. Bryant, *A New System*, 6:2–3.

63. The translation by J. Leitch was called *Introduction to a Scientific System of Mythology* (London: Longman, Brown, Green, and Longman, 1844). Concerning Müller, see Josine H. Blok, "Quests for a Scientific Mythology: F. Creuzer and K. O. Müller on History and Myth," *History and Theory* 33, no. 4 (1994): 26–52; Josine H. Blok, "Romantische Poesie, Naturphilosophie, Construktion der Geschichte: K. O. Müller's Understanding of History and Myth," in *K. O. Müller und die antike Kultur*, ed. W. M. Calder III and R. Schlesier (Hildesheim: Georg Olms Verlag, 1998), 55–97.

64. In general, see George Stocking, *Victorian Anthropology* (New York: Simon and Schuster, 1991), 8–45. See also Georges Gusdorf, *Les sciences humaines et la pensée occidentale*, vol. 5, *Dieu, la nature, l'homme au siècle des lumières* (Paris: Payot, 1972), 168–189.

65. Jacques-Joseph Champollion, *Dictionnaire égyptien en écriture hiéroglyphique* (Paris: Firmin Didot, 1841). Especially concerning the subsequent priority disputes, see R. B. Parkinson, *Cracking Codes: The Rosetta Stone and Decipherment* (Berkeley: University of California Press, 1999), 25–45. See also David Gange, *Dialogues with the Dead: Egyptology in British Culture and Religion, 1822–1922* (Oxford: Oxford University Press, 20013), 1–120. On Champollion's discovery, see Alain Faure, *Champollion, le savant déchiffré* (Paris: Fayard, 2004), 413–468. See also Jean Leclant, "La modification d'un regard (1787–1826): du Voyage en Syrie et en Égypte de Volney au Louvre de Champollion," *Comptes rendus des séances de l'Académie des Inscriptions et Belles-Lettres* 131, no. 4 (1987): 709–729.

66. For an outline of Kircher's attempts, see Daniel Stolzenberg, "Theory and Practice in Athanasius Kircher's Translations of the Hieroglyphs," in *Philosophers and Hieroglyphs*, Lucia Morra, and Carla Bazzanella (Turin: Rosenberg and Sellier, 2003), 74–99.

67. Brian Young, "'The Lust of Empire and Religious Hate': Christianity, History, and India, 1790–1820," in *History, Religion, and Culture: British Intellectual History 1750–1950*, ed. Stefan Collini, Richard Whatmore, and Brian Young (Cambridge, UK: Cambridge University Press, 2000), 91–111. Here and below, for relevant background, see Edward Said, *Orientalism* (London: Routledge,

1978), chap. 1. See also Geoffrey P. Nash, "New Orientalisms for Old: Articulations of the East in Raymond Schwab, Edward Said, and Two Nineteenth-Century French Orientalists," in *Orientalism Revisited: Art, Land, and Voyage*, ed. Ian Richard Netton (London: Routledge, 2013), 87–97.

68. James Mill, *A History of British India*, 2nd ed. (London: Baldwin, Cradock, and Joy, 1820), 1:357, 365.

69. Bryant, *A New System*, 1:278n30.

70. For Smith's scheme, see Adam Smith, *Lectures on Jurisprudence*, ed. R. L. Meek, D. D. Raphael, and P. G. Stein (New York: Oxford University Press, 1978), 1:27–32. For Mill's, see John Stuart Mill, "Civilization," *London and Westminster Review* 3 (April 1836): 1–27.

71. See Georg Wilhelm Friedrich Hegel, *The Philosophy of History*, trans. J. Sibree, intro. C. J. Friedrich (New York: Dover Publications, 1956), 105–110. On this, see Shlomo Avineri, *Hegel's Theory of the Modern State* (Cambridge, UK: Cambridge University Press, 1974), 221–238.

72. For Comte's scheme, see (among other places) Auguste Comte, *Cours de philosophie positive* (Paris: Rouen Frères, 1830), 1:1–56. Concerning other stage theories, see José Ferrater Mora, *Cuatro visiones de la historia universal: San Agustín, Vico, Voltaire, Hegel* (Madrid: Alianza Editorial, 1982). See also Johan Galtung and Sohail Inayatullah, eds., *Macrohistory and Macrohistorians: Perspectives on Individual, Social, and Civilizational Change* (New York: Praeger, 1997), chap. 1, 1–10; and Leon Pompa, *Vico: A Study of the "New Science,"* 2nd ed. (Cambridge, UK: Cambridge University Press, 1990), chaps. 9–12.

73. In regard to this aspect, see Tessa Morrison, *Isaac Newton's Temple of Solomon and His Reconstruction of Sacred Architecture* (Basel: Springer, 2011), 13–28; and Jed Z. Buchwald and Mordechai Feingold, *Newton and the Origin of Civilization* (Princeton: Princeton University Press, 2013), 164–221.

74. According to Hindu mythology, Vaivasvata, also known as Sraddhadeva or Satyavrata, was the king of Dravida before the great flood.

75. "This theology affords a remarkable instance of that progress in exaggeration and flattery which I have described as the genius of rude religion. As the Hindus, instead of selecting one god, to whom they assigned all power in heaven and in earth, distributed the creation and administration of the universe among three divinities, they divided themselves into sects; and some attached themselves more particularly to one deity, some to another." James Mill, *The History of British India*, 3rd ed., 6 vols. (London: Baldwin, Cradock, and Joy, 1826), 1:314.

76. See Edward Gibbon, *The History of the Decline and Fall of the Roman Empire*, 6 vols. (London: Strahan and Cadell, 1776–1789), especially vol. 1, chaps. 15–16. See also J. G. A. Pocock, *Barbarism and Religion, Volume 4, Barbarians, Savages,*

and Empires (Cambridge, UK: Cambridge University Press, 2005), 331–339. Concerning Gibbon's reception, see David Womersley, *Gibbon and the 'Watchmen of the Holy City': The Historian and His Reputation, 1776–1815* (Oxford: Oxford University Press, 2002).

77. On the realities of empire, see C. A. Bayly, *Imperial Meridian: The British Empire and the World 1780–1830* (London: Routledge, 1988), especially 133–163. For an illuminating introduction to the background of imperial thought, see David Armitage, *The Ideological Origins of the British Empire* (Cambridge, UK: Cambridge University Press, 2000), especially 1–23.

78. H. S. Milford, ed., *Oxford Standard Authors Poetical Works of Cowper*, rev. Norma Russell (Oxford: Oxford University Press, 1967), 184. Concerning this poem and other aspects of Cowper's works, see Dustin Griffin, *Patriotism and Poetry in Eighteenth-Century Britain* (Cambridge, UK: Cambridge University Press, 2002), 256–259.

79. For a record of the event, see Oliver, *An Account of the Religious Houses*, 13. For some of the larger issues here I refer to B. Dooley, "Victorian Physico-Theology in George Boole," in *Physico-Theology in England and on the European Continent (1650–c. 1750)*, ed. Ann Blair (forthcoming).

80. See, among others, Steven Dick, *Plurality of Worlds: The Origins of the Extraterrestrial Life Debate from Democritus to Kant* (Cambridge, UK: Cambridge University Press, 1982), 142–175.

81. William Wordsworth, *The Major Works*, ed. Stephen Gill (Oxford: Oxford University Press, 2000), 92. For other literary references, see Karl S. Guthke, "Nightmare and Utopia: Extraterrestrial Worlds from Galileo to Goethe," *Early Science and Medicine* 8, no. 3 (2003): 173–195.

82. Michael J. Crowe, *The Extraterrestrial Life Debate, 1750–1900* (Mineola, NY: Dover Editions, 1999), 210.

83. Thomas Dick, *Celestial Scenery: or, the Wonders of the Planetary System Displayed* (London: Thomas Ward, 1838), 397ff.

84. John Pringle Nichol, *The Phenomena and Order of the Solar System* (Edinburgh: W. Tait, 1838), 118.

85. Henry Lord Brougham, *Speeches of Henry Lord Brougham upon Questions Relating to Public Rights, Duties, and Interests*, 4 vols. (Philadelphia: Lea and Blanchard, 1841), 2:136. On the issues of censorship and freedom of expression in this period, see Joss Marsh, *Word Crimes: Blasphemy, Culture, and Literature in Nineteenth-Century England* (Chicago: University of Chicago Press, 1998), 78–126.

86. On Milton's 1644 polemic and its context, see Randy Robertson, *Censorship and Conflict in Seventeenth-Century England: The Subtle Art of Division* (University Park: Pennsylvania State University Press, 2009), 100–129.

87. On the religious landscape in these times, see Boyd Hilton, *The Age of Atonement: The Influence of Evangelicalism on Social and Economic Thought, 1785–1865* (Oxford: Clarendon Press, 1986), 36–70. On Evangelicalism in England, see Doreen Rosman, *Evangelicals and Culture,* 2nd ed. (Eugene, OR: Pickwick, 2012), 14–30. See also, Denys Leighton, *The Greenian Moment: T. H. Green, Religion, and Political Argument in Victorian Britain* (Exeter: Imprint Academic, 2004), especially chap. 4.

88. Steven A. Jauss, *Sermons, Chiefly on Particular Occasions* (London: Longman, Hurst, Rees, Orme, and Brown, 1814), 1:4. See also Steven A. Jauss, "Associationism and Taste Theory in Archibald Alison's Essays," *Journal of Aesthetics and Art Criticism* 64, no. 4 (2006): 415–428.

89. *Christian Benevolence Enforced: In a Sermon Preached in the Parish Church of St. Martin, Leicester, on Sunday October 3d, 1802* (London: T. Bensley [printer], 1802); *Devout Observance of the Sabbath Enforced: In a Short Address to the Parishioners of St. Martin's and All Saints'* (Leicester: T. Cooke [printer], 1813).

90. Thomas S. Kidd, *The Great Awakening: The Roots of Evangelical Christianity in Colonial America* (New Haven, CT: Yale University Press, 2009), 13–23; Joseph A. Conforti, *Jonathan Edwards, Religious Tradition, and American Culture* (Chapel Hill: University of North Carolina Press, 1995), 11–35.

91. Everest, "The Home-Side of a Scientific Mind," 335.

92. Ibid., 454. See also Cohen, *Equations from God,* chap. 3.

93. For the case of Barrow, see Antoni Malet, "Isaac Barrow on the Mathematization of Nature: Theological Voluntarism and the Rise of Geometrical Optics," *Journal of the History of Ideas* 58, no. 2 (1997): 265–287. The wider issues relating to the naturalization of the human and the humanization of nature are discussed in Stephen Gaukroger, *The Natural and the Human: Science and the Shaping of Modernity, 1739–1841* (Oxford: Oxford University Press, 2016), 1–16.

94. On this episode, see Robert Gray, *The Factory Question and Industrial England, 1830–1860* (Cambridge, UK: Cambridge University Press, 1996), 48–58, 190–212; Gary S. Cross, *A Quest for Time: The Reduction of Work in Britain and France, 1840–1940* (Berkeley: University of California Press, 1989), chaps. 2 and 4; Ann Provost Robson, *On Higher Than Commercial Grounds: The Factory Controversy, 1830–1853* (New York: Garland, 1985); Stewart Weaver, "The Political Ideology of Short Time: England, 1820–1850," in *Workflow: An International History,* ed. Gary Cross (Philadelphia: Temple University Press, 1989), 77–102.

95. Thomas Babington Macaulay, *Speeches of the Right Honorable T. B. Macaulay, M.P., Corrected by Himself* (London: Longman, Brown, Green, and Longmans, 1854), 451–452.

96. For relevant earlier studies, see Albert Larking, *History of the Early Closing Association to 1864* (London: Unwin Bros., 1914); Simon Rottenberg, "Legislated

Early Shop Closing in Britain," *Journal of Law and Economy* 4 (October 1961): 118–130; and Michael Winstanley, *The Shopkeeper's World, 1830–1914* (Manchester: Manchester University Press, 1983), 95–100.

97. Samuel Carter Hall, *Retrospect of a Long Life: From 1815 to 1883* (London: R. Bentley, 1883), 1:435.

98. Giovanni Federico, "The Corn Laws in Continental Perspective," *European Review of Economic History* 16, no. 2 (2012): 166–187; Anthony Howe, *Free Trade and Liberal England, 1846–1946* (Oxford: Oxford University Press, 1997), chaps. 1–3; Cheryl Schonhardt-Bailey, *From the Corn Laws to Free Trade: Interests, Ideas, and Institutions in Historical Perspective* (Cambridge, MA: MIT Press, 2006), 227–262; Andrew Marrison, ed., *Free Trade and Its Reception 1815–1960: Freedom and Trade: Volume 1* (London: Routledge 1998), chaps. 1, 2, 4, 13.

99. For excellent reflections on the context, see Saville, *1848: The British State and the Chartist Movement*, 80–101. See also Jean-Luc Mayaud, ed., *1848: Actes du colloque international du cent cinquantenaire, tenu à l'assemblée nationale à Paris, les 23–25 février 1998* (Paris: Créaphis, 2002).

100. Roberts, *Making English Morals,* chap. 4; Dominic Erdozain, *The Problem of Pleasure: Sport, Recreation, and the Crisis of Victorian Religion* (Woodbridge, UK: Boydell Press, 2010), chaps. 1–3.

101. For a discussion of the religious aspects, see Samuel Martin, *Serpents in Hedges: A Plea for Moderation in the Hours Employed in Business* (London: Ward and Co., 1850), 18.

102. Thomas Davies, *Prize Essay on the Evils Which Are Produced by Late Hours of Business, and on the Benefits Which Would Attend Their Abridgement* (London: James Nisbet and Company, 1843).

103. Cited in David W. Bebbington, Kenneth Dix, and Alan Ruston, eds., *Protestant Nonconformist Texts*, vol. 3, *The Nineteenth Century* (Eugene, OR: Wipf and Stock, 2006), 225.

104. Edward Baines, *The Social, Educational, and Religious State of the Manufacturing Districts* (London: Simpkin, Marshall, and Co., 1843), iv.

105. Concerning the issues below, see Paul Forster, "Kant, Boole, and Peirce's Early Metaphysics," *Synthese* 113 (1997): 43–70.

106. Edward Hyde, *The History of the Rebellion and Civil Wars in England*, 3 vols. (Oxford: printed at the theater, 1702–1704); Voltaire, *Le Siècle de Louis XIV* (Berlin: De Francheville, 1750).

107. Friedrich Schiller, *The History of the Thirty Years' War in Germany*, trans. M. Duncan (London: W. Simkin and R. Marshall, 1828), vol. 1, book 1, introduction.

108. I translate directly from Barthold Georg Niebuhr, *Römische Geschichte*, vol. 1 (Berlin: G. Reimer, 1828), x, rather than relying on *History of Rome*, trans. Julius Charles Hare and Connop Thirlwall (Cambridge, UK: John Taylor, 1828), 1:8.

109. Johann Melchior Hartmann and Johann David Ludewig Hess, *Allgemeines Register über die Göttingischen gelehrten Anzeigen von 1783 bis 1822* (Göttingen: Friedrich Ernst Huth, 1829), 3:479–484.

110. Alexander Fraser Tytler, *Universal History, From the Creation of the World to the Beginning of the Eighteenth Century*, 6 vols. (London: John Murray, 1834).

111. William Cooke Taylor, *Factories and the Factory System: From Parliamentary Documents and Personal Examination* (London: J. How, 1844), iii, 108, 115.

112. Henry Hart Milman, "Preface by the Editor," in *The History of the Decline and Fall of the Roman Empire*, by Edward Gibbon (Paris: Baudry, 1840), 1:xv.

113. Compare G. B. Airy, "Account of Some Circumstances Historically Connected with the Discovery of the Planet Exterior to Uranus," *Memoirs of the Royal Astronomical Society* 16 (1847): 385–414; D. W. Hughes, "J. C. Adams, Cambridge, and Neptune," *Notes and Records of the Royal Society* 50, no. 2 (1996): 245–248; Nicholas Kollerstrom, "An Hiatus in History: The British Claim for Neptune's Co-Prediction, 1845–46: Parts 1 and 2," *History of Science* 44 (2006): 3–28, 349–371.

114. Julie A. Reuben, *The Making of the Modern University: Intellectual Transformation and the Marginalization of Morality* (Chicago: University of Chicago Press, 1997), 21–22. For a careful comparison of Wayland's and Paley's works, see Donald E. Frey, *America's Economic Moralists: A History of Rival Ethics and Economics* (Albany, NY: SUNY Press, 2009), 42–45.

115. I will return to this theme in the context of the lecture "On Education." For the time being, see Lorraine Daston and Fernando Vidal, eds., *The Moral Authority of Nature* (Chicago: University of Chicago Press, 2003), especially editors' introduction; and Gaukroger, *The Natural and the Human*, 17–69.

116. On this aspect of Smith, see Jack Russell Weinstein, *Adam Smith's Pluralism: Rationality, Education, and the Moral Sentiments* (New Haven, CT: Yale University Press, 2013); David McNally, *Political Economy and the Rise of Capitalism: A Reinterpretation* (Berkeley: University of California Press, 1988), 228–250.

117. In general, see Chenxi Tang, *The Geographic Imagination of Modernity: Geography, Literature, and Philosophy in German Romanticism* (Stanford, CA: Stanford University Press, 2008), 25–55. See also Nicolas Pethes, *Zöglinge der Natur. Der literarische Menschenversuch des 18. Jahrhunderts* (Göttingen: Wallstein, 2007), chap. 4.

118. Concerning this theme, see Denis Lawton and Peter Gordon, *A History of Western Educational Ideas* (London: Woburn Press, 2002), especially chaps. 8–10, with bibliography.

119. Ian D. C. Newbould, "The Whigs, the Church, and Education, 1839," *Journal of British Studies* 26, no. 3 (1987): 332–346. See also John S. Hurt, *Education in Evolution: Church, State, Society, and Popular Education, 1800–1870* (London: Hart-Davis, 1971); David Phillips, *The German Example: English Interest in Educational Provision in Germany since 1800* (London: Bloomsbury Publishing, 2011); James Van Horn Melton, *Absolutism and the Eighteenth-Century Origins of Compulsory Schooling in Prussia and Austria* (Cambridge, UK: Cambridge University Press, 1988). On the Massachusetts system, see Susan Lynne Porter, ed., *Women of the Commonwealth: Work, Family, and Social Change in Nineteenth-Century Massachusetts* (Amherst: University of Massachusetts Press, 1996), 17–62.

120. Karl Marx and Friedrich Engels, *The Communist Manifesto: A Modern Edition*, ed. and intro. Eric Hobsbawm (London: Verso Books, 2012), 33.

121. Dieter Dowe, Heinz-Gerhard Haupt, Dieter Langewiesche, and Jonathan Sperber, *Europe in 1848: Revolution and Reform*, trans. David Higgins (New York: Berghahn Books, 2001). See also Jonathan Sperber, *The European Revolutions, 1848–1851* (Cambridge, UK: Cambridge University Press, 2005). On Britain, see Saville, *1848: The British State and the Chartist Movement*, 200–229.

122. Pestalozzi's ideas were diffused in Britain in part by works such as E. Biber, *Henry Pestalozzi and His Plan of Education* (London: Sherwood, Gilbert, and Piper, 1833). Concerning this strain of pedagogical thinking, see Käte Silber, *Pestalozzi: The Man and His Work* (London: Routledge and Kegan Paul, 1960). On the historiography of childhood, see Hugh Cunningham, "Histories of Childhood," *American Historical Review* 103, no. 4 (1998): 1195–1208.

123. On this aspect, see Dayna W. Murphree, "James Mill and Dugald Stewart on Mind and Education" (PhD diss., Virginia Polytechnic Institute, 2014), 313.

124. Consider Susan E. Cooperstein and Elizabeth Kocevar-Weidinger, "Beyond Active Learning: A Constructivist Approach to Learning," *Reference Services Review* 32, no. 2 (2004): 141–148.

125. Murphree, "James Mill and Dugald Stewart," 10.

126. See Thomas Chalmers, preface to *Lectures on Ethics by Thomas Brown* (Edinburgh: William Tait, 1846), xviii.

127. For a good discussion of the whole problem, see Thomas Dixon, *From Passions to Emotions: The Creation of a Secular Psychological Category* (Cambridge, UK: Cambridge University Press, 2003), chaps. 1–3. See also Charles L. Griswold, *Adam Smith and the Virtues of Enlightenment* (Cambridge, UK: Cambridge University Press, 1999); J. L. Mackie, *Ethics: Inventing Right and Wrong* (London: Pelican Books, 1977), especially chap. 8.

128. John Locke, *An Essay concerning Human Understanding*, ed. Roger Woolhouse (New York: Penguin Books, 1997), 357.

129. Johann Heinrich Pestalozzi, "Die Abendstunde eines Einsiedlers," *Ephemeriden der Menschheit* 1, no. 5 (May 1780), translated in Joseph Payne, *Pestalozzi; The Influence of His Principles and Practice on Elementary Education* (New York: E. Steiger, 1871), 9.

130. John Matteson, *Eden's Outcasts: The Story of Louisa May Alcott and Her Father* (New York: W. W. Norton, 2009), chap. 3.

131. Payne, *Pestalozzi*, 9.

132. John R. Davis, *The Victorians and Germany* (Oxford: Peter Lang, 2007), especially chap. 2.

133. Republished in William Whewell, *On the Principles of English University Education* (London: John W. Parker, 1838), 139.

134. George Boole, *The Mathematical Analysis of Logic* (London: George Bell, 1847), 4.

135. Compare Thomas Chalmers, *On the Power, Wisdom, and Goodness of God, as Manifested in the Adaptation of External Nature to the Moral and Intellectual Constitution of Man* (London: William Pickering, 1834), vol. 2, part II, chap. 4.

136. For an explanation of the religious environment, see W. Ralls, "The Papal Aggression of 1850: A Study in Victorian Anti-Catholicism," *Church History* 43 (1974): 242–256; Emmet Larkin, *The Making of the Roman Catholic Church in Ireland, 1850–1860* (Chapel Hill: University of North Carolina Press, 1980), 27–57.

137. Cohen, *Equations from God*, 83–84; Desmond MacHale, *George Boole: His Life and Work* (Dublin: Boole Press, 1985), chap. 6.

138. For a detailed analysis, see Jeffrey A. Auerbach, *The Great Exhibition of 1851: A Nation on Display* (New Haven, CT: Yale University Press, 1999).

139. *Economist* (London) 9 (1851): 4–6.

140. Cork, University College Cork Library, Special Collections, Boole Papers, BP/1/70, dated June 28, 1851, to his sister Maryann.

141. Concerning this theme, here and below, see Robert A. Nisbet, *History of the Idea of Progress* (New Brunswick, NJ: Transaction Publishers, 1994), especially 171–178, also noting the persistence of the concept to the late twentieth century. For a useful nineteenth-century development of the theme, see J. B. Bury, *The Idea of Progress; An Inquiry into Its Origin and Growth* (London, Macmillan and Co., 1921), 290–333.

142. James B. Conant, *On Understanding Science: An Historical Approach* (New Haven CT: Yale University Press, 1947), 5.

143. Jean-Antoine-Nicolas de Caritat marquis de Condorcet, *Outlines of an Historical View of the Human Mind* (London: J. Johnson, 1795), 319, 371.

144. Auguste Comte, *Cour de philosophie positive*, vol. 1, *Les préliminaires généraux et la philosophie mathématique* (Paris: Bachelier, 1830), 5. Compare William Adam, *An Inquiry into the Theories of History with Special Reference to the Principles of the Positive Philosophy* (London: W. H. Allen, 1862), 144.

145. John Stuart Mill, *System of Logic, Ratiocinative and Inductive* (London: John W. Parker, 1843), 2:292.

146. Letter to Augustus De Morgan, October 1850, cited in Cohen, *Equations from God*, 91n68. Concerning Boole and Kant, see Paul Forster, "Kant, Boole, and Peirce's Early Metaphysics," *Synthèse* 113, no. 1 (1997): 43–70.

147. Thomas Chalmers, *On the Power, Wisdom, and Goodness of God, as Manifested in the Adaptation of External Nature to the Moral and Intellectual Constitution of Man* (London: William Pickering, 1834). It appeared among the Bridgewater Treatises and comprised 2 volumes.

148. On the theme of the intelligibility of nature, see Peter Dear, *The Intelligibility of Nature: How Science Makes Sense of the World* (Chicago: University of Chicago Press, 2006).

149. Joel Mokyr and Cormac Ó Gráda, "Famine Disease and Famine Mortality: Lessons from Ireland, 1845–1850," *Famine Demography: Perspectives from the Past and Present*, ed. Tim Dyson (Oxford: Oxford University Press, 2002), 19–43.

150. A. J. P. Taylor, *The Struggle for Mastery in Europe: 1848–1918* (Oxford: Clarendon Press, 1954), 62–82. Concerning the war's unpopularity as well as the unprecedented media coverage, see Stefanie Markovits, *The Crimean War in the British Imagination* (Cambridge, UK: Cambridge University Press, 2009), 12–62.

151. On the dating of this sonnet to October–November 1655, weeks after the Swedish occupation of Warsaw, see John Thomas Shawcross, ed., *The Complete English Poetry of John Milton* (New York: New York University Press, 1963), 564.

152. Malcolm Andrews, *Charles Dickens and His Performing Selves: Dickens and the Public Readings* (Oxford: Oxford University Press, 2007), 271. On Benson: R. C. Cox, *Oxford Dictionary of National Biography* (Oxford: Oxford University Press, 2004), accessed April 14, 2017, http://www.oxforddnb.com/public/index.html. See also Joan Rockley, *Antiquarians and Archaeology in Nineteenth-Century Cork* (Oxford: Archaeopress, 2008), 59.

153. On the political issue, older interpretations such as Llewellyn Woodward, *The Age of Reform: 1815–1870* (New York: Doubleday, 1962) are now viewed in the light of, for instance, Richard Price, *British Society 1680–1880: Dynamism, Containment, and Change* (Cambridge, UK: Cambridge University Press, 1999), 264–291, relevant bibliography. In a broader comparative perspective,

see Jörg Neuheiser, *Krone, Kirche und Verfassung: Konservatismus in den englischen Unterschichten 1815–1867* (Göttingen:Vandenhoeck and Ruprecht, 2010).

154. Elizabeth Neswald, "The Benefits of a Mechanics' Institute and the Blessing of Temperance": Science and Temperance in 1840s' Ireland," *Social History of Alcohol and Drugs* 22 (2008): 209–227.

155. Jim Kemmy, "James Roche," *Old Limerick Journal* 25 (1989): 51–53; Richard Hayes, *Ireland and Irishmen in the French Revolution* (London: Ernest Benn Ltd., 1932), 190; Richard Hayes, *Biographical Dictionary of Irishmen in France* (Dublin: M. H. Gill and Son, 1949), 275–276.

156. Dorinda Outram, "Uncertain Legislator: Georges Cuvier's Laws of Nature in Their Intellectual Context," *Journal of the History of Biology* 19 (1986): 323–368.

157. M. MacSweeney and J. Reilly, "The Cork Cuvierian Society," *Journal of the Cork Historical and Archaeological Society* 63 (1958): 9–14.

158. University College Cork Library, Special Collections, ms. U.221 A, 234, entry dated March 7, 1855.

159. Ibid., 207, 234.

160. Ibid., 222, entry dated May 11, 1853.

161. Ibid., 224, entry dated October 5, 1853.

162. Ibid., 219, entry dated February 1853.

163. Ibid., 221, entry dated April 27, 1853.

164. Ibid., 230, entry dated May 3, 1854.

165. John Francis Maguire, *The Industrial Movement in Ireland as Illustrated by the National Exhibition 1852* (Dublin: J. M. Glashan, 1853), 3. See also A. C. Davies, "The First Irish Industrial Exhibition: Cork 1852," *Irish Economic and Social History* 2 (1975): 46–59. On the relation between this and the subsequent Dublin exhibition, see John Turpin, "Exhibitions of Art and Industries in Victorian Ireland: Part I: The Irish Arts and Industries Exhibition Movement 1834–1864," *Dublin Historical Record* 35 (1981): 2–13.

166. See Richard J. Kelly, "Queen Victoria and the Irish Industrial Exhibition of 1853," *Studies in Victorian Culture* 8 (2010): 3–30.

167. Joseph John Lee, *The Modernisation of Irish Society 1848–1918: From the Great Famine to Independent Ireland*, 2nd ed. (London: Gill and Macmillan Ltd., 2008), chap. 1. See also Cormac Ó Gráda, *Ireland before and after the Famine: Explorations in Economic History, 1800–1925*, rev. ed. (Manchester: Manchester University Press, 1993), chap. 2; Cormac Ó Gráda, *Ireland: A New Economic History, 1780–1939* (New York: Oxford University Press, 1994), chaps. 12–13. For the prefamine industrial situation, see Frank Geary, "Regional Industrial

Structure and Labour Force Decline in Ireland between 1841 and 1851," *Irish Historical Studies* 30 (1996): 167–194.

168. For excerpts from the 1851 census, see John Killen, ed., *The Famine Decade: Contemporary Accounts, 1841–51* (Belfast: Blackstaff Press, 1995), 247. See also *The Census of Ireland for the Year 1851*, 5 parts (Dublin: A. Thom and Sons, 1852–1856).

169. For a basic interpretation and bibliography, see Jürgen Osterhammel, *The Transformation of the World: A Global History of the Nineteenth Century* (Princeton, NJ: Princeton University Press, 2014), 241–256.

170. See J. P. McCarthy, "Dr. Richard Caulfield: Antiquarian, Scholar, and Academic Librarian," *Journal of the Cork Historical and Archaeological Society* 92 (1987): 1–23; Leaney, "Missionaries of Science."

171. For a detailed nearly contemporary testimony about this item, see by Eugene O'Curry, *Lectures on the Manuscript Materials of Ancient Irish History* (Dublin 1861), 196–200.

172. R. Caulfield, "Report on the Department of Antiquities," *Report of the Cuvierian Society for the Cultivation of the Sciences, for the Session 1854–55* (Cork: George Purcell and Co., 1855), 11–15.

173. T. S. Dunscombe, "Report on the Department of Art," *Report of the Cuvierian Society for the Cultivation of the Sciences, for the Session 1854–55* (Cork: George Purcell and Co., 1855), 15.

174. Charles G. Haines, "Report on the Department of Natural History," *Report of the Cuvierian Society for the Cultivation of the Sciences, for the Session 1854–55* (Cork: George Purcell and Co., 1855), 8–11.

175. George F. Shaw, "Report on the Physical and Experimental Sciences," *Report of the Cuvierian Society for the Cultivation of the Sciences, for the Session 1854–55* (Cork: George Purcell and Co., 1855), 7–8.

176. Anonymous, "Account of the Conversazione for the Working Classes on the Evening of Thursday, May 31st," *Report of the Cuvierian Society for the Cultivation of the Sciences, for the Session 1854–55* (Cork: George Purcell and Co., 1855), 16–19. See also Denis Gwynn, "Cork Cuvierian Society, 1849–1851," *Cork University Record* 23 (1951): 27–34.

177. Kane, *The Industrial Resources of Ireland*, 395.

178. My reflections here are much inspired by Ian Hacking, *The Social Construction of What?* (Cambridge, MA: Harvard University Press, 1999), chaps. 1–3.

179. For a particularly evocative account of this shift, see Michel Foucault, *The Birth of the Clinic: An Archaeology of Medical Perception circa 1780–circa 1830*, trans. A. M. Sheridan (London: Tavistock Publications, 1973), ix–xix. Combining historiography with social critique, see also Harold Perkin, *The Rise of Professional*

Society: England since 1880 (London: Routledge, 1990). For stimulating reflections, see Jean Boutier, Jean-Claude Passeron, and Jacques Revel, eds., *Qu'est-ce qu'une discipline?* (Paris: Enquête, Éditions de l'EPHESS, 2006).

180. The following is drawn from *Reports of Presidents of Queen's Colleges at Belfast, Cork, and Galway, 1849–50* (Dublin: H. M. Stationery Office, 1850), 33ff.

181. Adolphe Quetelet, *Facts, Laws and Phenomena of Natural Philosophy, Or, Summary of a Course of General Physics*, trans. Robert Wallace (Glasgow: P. Sinclair, 1835), 3.

182. On definitions of science, see Ziauddin Sardar, *Thomas Kuhn and the Science Wars* (London: Icon Books, 2000), 9.

183. Céran Lemonnier, *A Synopsis of Natural History: Embracing the Natural History of Animals, with Human and General Animal Physiology, Botany, Vegetable Physiology and Geology*, trans. Thomas Wyatt (Philadelphia: T. Wardle, 1839), title page.

184. George James Allman, *Introductory Lecture Delivered to the Students of the Natural History Class in the University of Edinburgh on the Opening of the Winter Session 1855* (Edinburgh: Adam and Charles Black, 1855), 4.

185. See Hilde de Ridder-Symoens, ed., *A History of the University in Europe*, vol. 2, *Universities in Early Modern Europe (1500–1800)* (Cambridge, UK: Cambridge University Press, 1996).

186. *Monthly Magazine or British Register* 51 (1821): 249.

187. For the classic study, see Morris Berman, "'Hegemony' and the Amateur Tradition in British Science," *Journal of Social History* 8 (1975): 30–50. For a recent critique, see Aileen Fyfe and Bernard Lightman, eds., *Science in the Marketplace: Nineteenth-Century Sites and Experiences* (Chicago: University of Chicago Press, 2007), 2–3. From a different perspective, see Joseph Ben-David, *The Scientist's Role in Society* (Englewood Cliffs, NJ: Prentice-Hall, 1971), chap. 7; on which, Liah Greenfeld, ed., *The Ideals of Joseph Ben-David: The Scientist's Role and Centers of Learning Revisited* (New Brunswick, NJ: Transaction Publishers, 2012).

188. For a discussion of this development, see R. Steven Turner, "The Growth of Professorial Research in Prussia, 1818 to 1848—Causes and Context," *Historical Studies in the Physical Sciences* 3 (1971): 137–182. For a particularly insightful work in regard to the subsequent historiography, see David Cahan, ed., *From Natural Philosophy to the Sciences: Writing the History of Nineteenth-Century Science* (Chicago: University of Chicago Press, 2003), chaps. 1–2, 10.

189. Max Weber, *The Protestant Ethic and the Spirit of Capitalism*, trans. Talcott Parsons, intro. Anthony Giddens (London: Allen and Unwin, 1976), 182.

190. Marie Boas Hall, *The Library and Archives of the Royal Society 1660–1990* (London: Royal Society, 1992), 37.

191. George Boole, *Selected Manuscripts on Logic and Its Philosophy*, ed. Ivor Grattan-Guinness and Gérard Bornet (Basel: Springer, 1997), 203–221.

192. Everest, "The Home-Side of a Scientific Mind," 105–114, 173–183, 327–336, 456–460.

CHAPTER 1

1. Originally published as George Boole, *An Address on the Genius and Discoveries of Sir Isaac Newton, Delivered on Thursday, February 5, 1835, at the Lincoln and Lincolnshire Mechanics Institution* (Lincoln: Gazette Office, printer, 1835). Boole, a member of the institution, gave this lecture at the presentation of a marble bust of Newton by Charles Anderson-Pelham, Baron Yarborough (and from 1837, first Earl).

2. There is a voluminous and sometimes controversial literature concerning Newton's early childhood and the possible influences on his future development. For a summary, see Richard S. Westfall, *Never at Rest: A Biography of Isaac Newton* (Cambridge, UK: Cambridge University Press, 1981), chap. 2. For a more recent work, see A. Rupert Hall, *Isaac Newton: Adventurer in Thought* (Oxford: Blackwell, 1992), chap. 1.

3. Much here relies on David Brewster, *The Life of Sir Isaac Newton* (London: J. Murray, 1831), chap. 1. This work was later reissued in an augmented edition as *Memoirs of the Life, Writings, and Discoveries of Sir Isaac Newton* (London: Hamilton, Adams and Co., 1855).

4. Costantino Grimaldi (1667–1750), member of the Accademia degli Investiganti in Naples, of whom the following work is referenced here: *Risposta alla terza lettera apologetica contra il Cartesio creduto da più d'Aristotle di Benedetto Aletino. Opera, in cui dimostrasi quanto salda, e pia sia la filosofia di Renato Delle Carte e perchè questo si debba stimare piu d'Aristotele* (Colonia [=Naples]: Sebastiano Hecht, 1703), 465.

5. The paper read at successive meetings of the Royal Society in the year 1675 is conventionally known as the "Hypothesis Explaining the Properties of Light," whereas the books of the Royal Society refer to a "manuscript of Mr. NEWTON, touching his theory of light and colours, containing partly an hypothesis to explain the properties of light discoursed of by him in his former papers," as recorded in Thomas Birch, *The History of the Royal Society* (London, 1757), 3:247–305. The full version was printed as Isaac Newton, *Opticks: or, a Treatise of the Reflexions, Refractions, Inflexions, and Colours of Light* (London: Smith and Walford, 1704). See Alan E. Shapiro, "Newton's Optics and Atomism," in

I. Bernard Cohen and George E. Smith eds., *The Cambridge Companion to Newton* (Cambridge, UK: Cambridge University Press, 2002), 227–255.

6. On this and many other aspects of Newton's theory, see Alan E. Shapiro, *Fits, Passions, and Paroxysms: Physics, Method, and Chemistry, and Newton's Theories of Coloured Bodies and Fits of Easy Reflections* (Cambridge, UK: Cambridge University Press, 1993), chap. 4.

7. On this discrepancy and its results, see Brewster, *The Life of Sir Isaac Newton*, chap. 11.

8. Isaac Newton, *Philosophiae Naturalis Principia Mathematica* (London: Joseph Streater for the Royal Society, 1687). A translation by A. Motte appeared as *The Mathematical Principles of Natural Philosophy* (London: B. Motte, 1729).

9. The apocryphal story was transmitted by Sir Walter Scott in the novel *The Antiquary* (London: Longman, Hurst, Rees, Orme, and Brown, 1816), vol. 2, chap. 1. It was picked up in Brewster, *The Life of Sir Isaac Newton*, chap. 13. The wording in Boole's version, however, is as close to Scott's as it is far from Brewster's, indicating the probable source is the former.

10. The translation from the original Latin inscription, significantly different from the one in Brewster, *Life of Sir Isaac Newton*, 325, appears to be Boole's own.

CHAPTER 2

1. From the manuscript in University College Cork, Boole Library, Special Collections, BP/1/270–1.

2. The Cuthites lived in 6th century BC Samaria, and the Pelasgians inhabited the Aegean area before the Greeks.

3. Compare Jacob Bryant, *A New System or an Analysis of Ancient Mythology* (London: J. Walker, 1807), 1:276ff.; Herodotus, *Histories*, 1:131.

4. Compare Plutarch, *Isis and Osiris*, in *Moralia*, vol. 5, trans. Frank Cole Babbitt (Cambridge, MA: Harvard University Press, 1936), sec. 46 (369E): "[Zoroaster] called the one Oromazes and the other Areimanius; and he further declared that among all the things perceptible to the senses, Oromazes may best be compared to light, and Areimanius, conversely, to darkness and ignorance, and midway between the two is Mithras: for this reason the Persians give to Mithras the name of 'Mediator.'"

5. Herodotus, *Histories*, 1:131. The quote is rather liberally excerpted from the translation by William Beloe, originally published in 4 volumes (London: Leigh and Sotheby, 1791), 1:104. Newton referenced the same passage in *The Chronology of Ancient Kingdoms Amended* (London: J. Tonson, J. Osborn, and T. Longman, 1728), chap. 2, 220.

6. As Bryant (*A New System*, 1:94) commented, "How little we know of Druidical worship, either in respect to its essence or its origin." Yet Vincenzo Bellini's Druidical opera *Norma*, with a libretto by Felice Romani, in turn partly based on Vicomte Chateaubriand's novel *Les martyrs* (1809), was a hit in London in 1833.

7. All from Plutarch, *Isis and Osiris*, sec. 47 (370A).

8. The ichneumon is an Egyptian mongoose, and in mythology, the animal into which Ra metamorphoses in order to fight the evil god-snake Apopis.

9. The word "that" in this verse was "who" in the original poem by John Milton, *Paradise Lost* (1:477) in vol. 1 of *Poetical Works of John Milton*, ed. Helen Darbishire (Oxford: Clarendon Press, 1963), 17.

10. The four next lines are omitted.

11. Diodorus Siculus, *Library of History*, trans. C. H. Oldfather (Cambridge, MA: Harvard University Press, 1933), 1:11.

12. Ibid., 1:14.

13. Diodorus Siculus, *Library of History*, 5:4.

14. Bryant, *A New System*, 3:182.

15. Genesis 8:20 and 9:20, respectively.

16. Plutarch, *Isis and Osiris*, sec. 21 (359C–359E).

17. Here at "allotted to the priests," UCC, MS BP/1/270 breaks off, and MS BP/1/271 begins.

18. On the parallelism between Hinduism and New World religions, see John Delafield, *An Inquiry into the Origin of the Antiquities of America: With an Appendix Containing Notes and "A View of the Causes of the Superiority of the Men of the Northern over Those of the Southern Hemisphere" by James Lakey* (London: Longman, Rees, Orme, Brown, Green, and Longman, 1839), especially 31, citing a *Rees's Cyclopedia* article suggesting that "the Peruvian worship was that of Vishnu, when he appeared under the form of Chrishna, or the sun: whilst the sanguinary worship of the Mexicans is analogous to that of Siva, when he takes the character of the Stygian Jupiter." On cannibalism, see Bryant, *A New System*, 6:311.

19. William Stukeley, *Stonehenge, a Temple Restor'd to the British Druids* (London: W. Innys and R. Manby, 1740); Richard Colt Hoare, *The History of Ancient Wiltshire*, 2 vols. (London: William Miller, 1810–1821); John Bathurst Deane, *The Worship of the Serpent* (London: Gilbert and Rivington, 1833). The Druid connection with Stonehenge was debunked by T. D. Kendrick, *The Druids: A Study in Keltic Prehistory* (London: Methuen, 1927).

20. Henry Bliss [Nicholas Thirning Moile, pseud.], *State Trials, Specimen of a New Edition* (London: Simpkin, Marshall, and Co., 1838). The work, consisting of

fanciful poetry on the trials of Anne Ayliffe for heresy, Sir William Stanley for high treason, and Mary Queen of Scots (for unspecified crimes) met with an enthusiastic response in the *Monthly Magazine* 1 (1839): 197–200.

21. For the then-current account of these caverns, see Captain Robert Elliot, *View in the East, with Historical and Descriptive Illustrations*, vol. 1 (London: H. Fisher, 1833).

22. Antoine-Yves Goguet, *The Origin of Laws, Arts, and Sciences, and Their Progress among the Most Ancient Nations* [*De l'origine des loix, des arts, et des sciences; et de leurs progrès chez les anciens peuples*], trans. Robert Henry (Edinburgh: George Robinson, 1775), book 2, chap. 3, 1:132–139.

23. Paraphrased from Edward Gibbon, *The History of the Decline and Fall of the Roman Empire*, 6 vols. (London: Strahan and Cadell, 1776–1789), 1:29.

24. From William Cowper's poem "The Task."

25. James Mill, *A History of British India*, 2nd ed. (London: Baldwin, Cradock, and Joy, 1820), 1:368, in which the quote is actually cited from Francis Buchanan, *A Journey from Madras through the countries of Mysore, Canara and Malabar* (London: T. Cadell and W. Davies, 1807), 1:167.

26. Ibid., 1:369, in which the quote is actually cited from John Fryer, *A New Account of East-India and Persia* (London: Printed by R.R. for R. Chiswell, 1698), chap. 5, sec. 3.

CHAPTER 3

1. Original title: "On the Question: Are the Planets Inhabited? and, On the Stability of the Planetary System." Manuscript in University College Cork, Boole Library, Special Collections, BP/1/272.

2. Compare Immanuel Kant, *Critique of Pure Reason*, trans. and ed. Paul Guyer and Allen W. Wood (Cambridge, UK: Cambridge University Press, 1998), 578: "If, then, neither the concept of things in general nor the experience of any existence in general can achieve what is required, then one means is still left: to see whether a determinate experience, that of the things in the present world, their constitution and order, yields a ground of proof that could help us to acquire a certain conviction of the existence of a highest being. Such a proof we would call the physico-theological proof. If this too should be impossible, then no satisfactory proof from speculative reason for the existence of a being that corresponds to our transcendental ideas is possible at all." The original phrase is in Immanuel Kant, *Kritik der reinen Vernunft, Zweite hin und wieder verbesserte Auflage* (Riga: Johann Friedrich Hartnoch, 1787), 649.

3. The reference is to John Pringle Nichol, *The Phenomena and Order of the Solar System* (Edinburgh: W. Tait, 1838).

4. The reference is to Peter Leonard, *The Western Coast of Africa: Journal of an Officer under Captain Owen. Records of a Voyage in the Ship Dryad in 1830, 1831, and 1832* (London: E. C. Mielke, 1833).

5. Georges Cuvier, *Recherches sur les ossemens fossiles de quadrupèdes, où l'on rétablit les caractères de plusieurs espèces d'animaux que les révolutions du globe paroissent avoir détruites*, 4 vols. (Paris: Deterville, 1812).

6. Concerning the discoveries of Christian Gottfried Ehrenberg that are at issue here, see Frederick B. Churchill, "The Guts of the Matter. Infusoria from Ehrenberg to Bütschli: 1838–1876," *Journal of the History of Biology* 22, no. 2 (1989): 189–213.

7. Relevant here is Kenneth F. Schaffner, *Nineteenth-Century Aether Theories* (Oxford: Pergamon Press, 1972), chap. 1.

8. Encke's Comet, the next discovered after Halley's, derived its name from Johann Franz Encke, who calculated its orbit in 1819.

9. Ovid, *Metamorphoses*, 1:254–258.

10. 2 Peter 3:10.

CHAPTER 4

1. From the manuscript in the Royal Society Library, mss. 782, add. box 8.

2. Concerning some of these authors, see the introduction. Richard Whately was a prolific theologian and moralist who became the Church of Ireland archbishop of Dublin. For Thomas Chalmers, see the lecture "The Right Use of Leisure." Isaac Barrow was Isaac Newton's predecessor in the Lucasian Professorship at Oxford. Brook Taylor, a contemporary of Newton, was a mathematician who also wrote a *Contemplatio Philosophica*, posthumously published by his grandson in 1793.

3. John Bird Sumner became the archbishop of Canterbury, and is best known for his *Treatise on the Records of Creation and the Moral Attributes of the Creator* (London, 1816) and *The Evidence of Christianity Derived from Its Nature and Reception* (London, 1821).

4. Thomas Arnold (1795–1842), better known, at the time, for religious works such as *Sermons: Christian Life, Its Hopes, Fears and Close* (London: Fellowes, 1842) than for his unfinished *History of Rome*, 3 vols. (London: Fellowes, 1838–1842) or his *Introductory Lectures on Modern History* (London: Longmans, Green & Co, 1842).

5. The reference is to John Henry Newman, whose *Essay on the Development of Christian Doctrine* (London: James Toovey, 1845) went through several editions.

6. William Ellery Channing (1780–1842).

CHAPTER 5

1. Originally published as George Boole, *The Right Use of Leisure: An Address, Delivered before the Members of the Lincoln Early Closing Association, February 9th, 1847* (London: J. Nisbet and Company, 1847).

2. Thomas Chalmers, *On the Power, Wisdom, and Goodness of God, as Manifested in the Adaptation of External Nature to the Moral and Intellectual Constitution of Man* (London: William Pickering, 1834), 1:193: "The views contemplated through the medium of observation, are found, not only to have a justness in them, but to have a grace and a grandeur in them, far above all the visions which are contemplated through the medium of fancy, or which ever regaled the fondest enthusiast in the enraptured walks of speculation and poetry. But the toils of investigation must be endured first, that the grace and the grandeur might be enjoyed afterwards. The same is true of science in all its departments, not of simple and sublime astronomy alone, but throughout of terrestrial physics; and most of all in chemistry, where the internal processes of actual and ascertained Nature are found to possess a beauty, which far surpasses the crude though specious plausibilities of other days. We perceive in this too, a fine adaptation of the external world to the faculties of man; a happy ordination of Nature by which the labour of the spirit is made to precede the luxury of the spirit, or every disciple of science must strenuously labour in the investigation of its truths ere he can luxuriate in the contemplation of its beauties."

3. Relevant here is Bruce Haley, *The Healthy Body and Victorian Culture* (Cambridge, MA: Harvard University Press, 1978). The "mens sana in corpore sano" concept was most notably explored by Andrew Combe, *The Principles of Physiology Applied to the Preservation of Health, and to the Improvement of Physical and Mental Education* (Edinburgh: Adam and Charles Black, 1834), which went through eleven editions already by 1842. See also Samuel Smiles, *Physical Education: or, the Nurture and Management of Children, Founded on the Study of Their Nature and Constitution* (London: Oliver and Boyd, Tweeddale Court, 1838).

4. Robert Southey, "My Days among the Dead Are Past," in *Poems of Robert Southey*, ed. Maurice Henry Fitzgerald (Oxford: Oxford University Press, 1909), 347.

5. The works mentioned here are Alexander Fraser Tytler, *Elements of General History, Ancient and Modern*, 2 vols. (Edinburgh: W. Creech, 1801); Arnold Hermann Ludwig Heeren, *Handbuch der Geschichte der Staaten des Alterthums* (Göttingen: J. G. Rosenbuch's Wittwe, 1799), translated by David Alphonso Talboys as *Manual of Ancient History* (Oxford: D. A. Talboys, 1829); William Cooke Taylor, *The Student's Manual of Ancient History* (London: John W. Parker, 1834); William Cooke Taylor, *The Student's Manual of Modern History* (London: John W. Parker, 1838).

6. Relevant volumes from among the total of 133 belonging to Dionysius Lardner's *Cabinet Cyclopædia* (London: Longman, etc., 1829–1846) would have been Walter Scott's *History of Scotland,* William Desborough Cooley's *History of Maritime Discovery,* James Mackintosh's *History of England,* Thomas Colley Grattan's *The History of the Netherlands,* and so forth. Among these volumes was also a one-volume reduction (1832) of Jean Charles Léonard Simonde de Sismondi's 16-volume *Histoire des républiques italiennes du Moyen Âge* (published from 1807 to 1818) with a preface by the author. His 31-volume *Histoire des Français* (Paris: Treuttel et Würz, 1821–1844) was not translated yet. In addition, here, George Lillie Craik et al., *Pictorial History of England* (London: Charles Knight and Co., 1838–1844); David Hume, *The History of England,* 8 vols. (London: T. Cadell, 1778).

7. Barthold Georg Niebuhr, *The History of Rome,* trans. Julius Charles Hare and Connop Thirlwall, 2 vols. (Cambridge, UK: John Taylor, 1828–1832); Connop Thirlwall, *History of Greece,* 8 vols. (London: Longman, Rees, Orme, Brown, Green, and Longman and John Taylor, 1835–1844); William Robertson, *The History of Scotland,* 2 vols. (London: A. Millar, 1759); William Robertson, *The History of America,* 2 vols. (London: W. Strahan, 1777).

8. Mentioned here are William Robertson, *The History of the Reign of the Emperor Charles V,* 7th ed., 4 vols. (London: W. Strahan, 1792); Friedrich Schiller, *The History of the Thirty Years' War in Germany,* trans. M. Duncan, 2 vols. (London: W. Simkin and R. Marshall, 1828); Edward Hyde, Earl of Clarendon, *The History of the Rebellion and Civil Wars in England,* 3 vols. (Oxford: Theater, printer, 1702–1704); and John Forster, James Mackintosh, John Macdiarmid, and Thomas Peregrine Courtenay, *Lives of Eminent British Statesmen,* 7 vols. (London: Longman, etc., 1831–1841). The work of Voltaire came out in a revised translation by a certain Mr. Chambaud with the title *The Age of Lewis XIV: A New Edition, Revised, and Considerably Augmented, by the Author* (London: R. and J. Dodsley, 1753). Jean-Henri Merle d'Aubigné, *Histoire de la Reformation en Europe au temps de Calvin,* 8 vols., 1862–1877, came out in an English translation by W. K. Kelly as *History of the Reformation in the Sixteenth Century,* rev. ed. (London: Whittaker, 1842).

9. References are to François Guizot, *General History of Civilisation in Europe, from the Fall of the Roman Empire to the French Revolution* (Oxford: Talboys, 1837); Robert Vaughan, *The Age of Great Cities; or, Modern Civilization Viewed in Its Relation to Intelligence, Morals, and Religion* (London: Jackson and Walford, 1843); William Cooke Taylor, *The Natural History of Society in the Barbarous and Civilized State: An Essay Towards Discovering the Origin and Course of Human Improvement,* 2 vols. (London: Longman, Orme, Brown, Green, and Longmans, 1840). The Taylor reference is an educated guess.

10. Apart from Newton, the figures remembered here are John Howard (1726–1790), prison reformer; Robert Leighton (1611–1684), archbishop of Glasgow

and, for a time, principal of the University of Edinburgh; and Sir Philip Sidney (1554–1586), literary and political figure.

11. At issue is the discovery of Neptune contested by Urbain Le Verrier and John Couch Adams in 1847; see the introduction.

12. The texts mentioned here include Golding Bird, *Elements of Natural Philosophy: Being an Experimental Introduction to the Study of the Physical Sciences* (London: John Churchill, 1839), based on lectures to medical students at Guy's Hospital; Thomas Young's lectures at the Royal Institution were first published in 1807 and republished, with notes and references by Philip Kelland, professor of mathematics at the University of Edinburgh, as *A Course of Lectures on Natural Philosophy and the Mechanical Arts, a New Ed., with References and Notes by Philip Kelland*, 2 vols. (London: Taylor and Walton, 1845); Carl Friedrich Peschel's 1842 *Lehrbuch der Physik* was translated by Ebenezer West as *Elements of Physics*, 3 vols. (London: Longman, Brown, Green, and Longmans, 1845–1846).

13. For the differences between William Paley and Francis Wayland, see the introduction.

14. The texts mentioned here include William Whewell, *The Elements of Morality: Including Polity*, 2 vols. (London: John Parker, 1845); James Mackintosh, *Dissertation on the Progress of Ethical Philosophy: Chiefly during the Seventeenth and Eighteenth Centuries* (Edinburgh: A. and C. Black, 1837); Thomas Brown, *Lectures on Ethics*, preface by Thomas Chalmers (Edinburgh: William Tait, 1846).

15. Jonathan Dymond, *Essays on the Principles of Morality, and on the Private and Political Rights and Obligations of Mankind*, 2 vols. (London: Hamilton, Adams, and Co, 1829). Dymond is also known for his treatises on pacifism.

16. Chalmers, *On the Power, Wisdom, and Goodness of God*; Ralph Wardlaw, *Christian Ethics: or, Moral Philosophy on the Principles of Divine Revelation* (London: Jackson and Walford, 1833).

17. Adam Clarke was a Methodist theologian chiefly known for a monumental commentary on the Bible; Sir William Jones was a pioneer in comparative linguistics.

18. Boole explores various belief systems in his lecture "On the Character and Origin of the Ancient Mythologies." Gnosticism was a second-century heretical movement within Christianity.

19. Richard Whately, *Essays on Some of the Peculiarities of the Christian Religion* (London: B. Fellowes, 1831), 368: "The charge of Priestcraft, so often brought indiscriminately against all religions, by those whose hostility is in fact directed against Christianity, falls entirely to the ground, when applied, not to the corruptions of the Romish Church (which certainly does lie open to the imputation), but to the religion of the Gospel, as founded on the writings of its promulgators. It is a religion which has no Priest on earth—no mortal Intercessor to stand

between God and his worshippers; but which teaches its votaries to apply, for themselves, to their great and divine High Priest, and to 'come boldly to the throne of grace, that they may find help in time of need.' Nor are the Christian Ministers appointed, as the infidel would insinuate, for the purpose of keeping the people in darkness, but expressly for the purpose of instructing them in their religion."

20. *The Iliad of Homer,* trans. Alexander Pope, book 6, line 181, in *Complete Poetical Works* (Boston: Houghton Mifflin, 1903). Compare the following: "As of the green leaves on a thick tree, some fall, and some grow" (Sirach 14:18).

21. Romans 14:7.

CHAPTER 6

1. Manuscript (holograph) in Royal Society Library, R.S. mss. 782, add. box 2; copy at University College Cork, Boole Library, Special Collections, BP/1/274.

2. Apart from the sources mentioned above, see Lucille M. Schultz, "Pestalozzi's Mark on Nineteenth-Century Composition Instruction: Ideas Not in Words, but in Things," *Rhetoric Review* 14 (1995): 23–43. See also Elizabeth Hoiem, "Object Lessons: Technologies of Education in British Literature, 1762–1851" (PhD diss., University of Illinois at Urbana-Champaign, 2013). Compare Eckhardt Fuchs, "Nature and Bildung: Pedagogical Naturalism in Nineteenth-Century Germany," in *The Moral Authority of Nature,* ed. Lorraine Daston and Fernando Vidal (Chicago: University of Chicago Press, 2003), 155–181.

3. Compare Thomas Wyse, *Education Reform, or, the Necessity of a National System of Education* (London: Longman, Rees, Orme, Brown, Green, and Longman, 1836), 144, unnumbered note: "Formerly, teachers would hardly allow maps in aid of books: now, books should hardly be allowed in aid of maps. The map should be the book."

4. Compare Johann Gottfried Herder, *Outlines of a Philosophy of the History of Man,* trans. T. Churchill (London: J. Johnson, 1803), 2:39: "The father of the World chose a more favourable spot for our origin. He placed the chief trunk of the mountains of the old world in the temperate zone, and the most cultivated nations dwell at its foot. Here he gave mankind a milder climate, and with it a gentler nature, and a more variegated place of education: thence he let them wander by degrees, strengthened and well instructed, into hotter and colder regions. There the primitive races could at first live in peace, then gradually draw off along the mountains and rivers, and become inured to ruder climates. Each cultivated its little circle, and enjoyed it, as if it had been the universe. Neither fortune nor misfortune spread itself so irresistibly wide, as if a probably higher chain of mountains under the equator had commanded the whole northern and southern world. Thus the Creator of the World has ever ordained things better

than we could have directed; and the irregular form of our Earth has effected an end, that greater regularity could never have accomplished."

5. Compare John Locke, *Essay concerning Human Understanding*, ed. John W. Yolton (London: Dent, 1961), vol. 1, book 2, chap. 1, 77: "Our observation, employed either about external sensible objects, or about the internal operations of our minds perceived and reflected on by ourselves, is that which supplies our understandings with all the materials of thinking. These two are the fountains of knowledge, from whence all the ideas we have, or can naturally have, do spring."

6. Compare E. Biber, *Henry Pestalozzi*, 217: "combining industry with education."

7. Added in pencil in Royal Society manuscript, after the phrase "instruction is," but not by Boole: "other things being equal."

8. Parentheses only in Royal Society manuscript.

9. Here there is a gap in the UCC manuscript.

10. Thomas Chalmers, preface to Thomas Brown, *Lectures on Ethics* (Edinburgh: William Tait, 1846).

CHAPTER 7

1. Originally published as George Boole, *The Claims of Science, Especially as Founded in Its Relations to Human Nature. Delivered in Queens College Cork at the Opening of the Third Session in October, 1851* (London: Tatloe, Walton, and Maberly, 1851).

2. Concerning the definitions of the term "science" to the mid-nineteenth century, see Katharine Park and Lorraine Daston, "Introduction: The Age of the New," *The Cambridge History of Science*, vol. 3, *Early Modern Science*, ed. Katharine Park and Lorraine Daston (Cambridge, UK: Cambridge University Press, 2006), 1–21. See also Andrew Cunningham and Perry Williams, "De-Centring the 'Big Picture': The Origins of Modern Science and the Modern Origins of Science," *British Journal for the History of Science* 26 (1993): 407–432. On the cross-cultural linguistic aspects, see Denise Phillips, *Acolytes of Nature: Defining Natural Science in Germany, 1770–1850* (Chicago: University of Chicago Press, 2012), 1–26.

3. Francis Bacon, *Novum Organum Scientiarum* (original ed. 1620), later published in translation in James Spedding, Robert Leslie Ellis, and Douglas Denon Heath, eds., *The Collected Works of Francis Bacon* (London: Longmans, 1857–1874), vol. 4.

4. The reference is to John Milton, *Paradise Lost* (5:175–179), in vol. 1 of *Poetical Works of John Milton*, ed. Helen Darbishire (Oxford: Clarendon Press, 1963), 104: "Moon, that now meet'st the orient Sun, now fliest, / With the fixed Stars, fixed

in their orb that flies; / And ye five other wandering Fires, that move / In mystic dance, not without song, resound / His praise who out of Darkness called up Light."

5. Concerning the theological picture here, see Frederick Gregory, "Intersections of Physical Science and Western Religion in the Nineteenth and Twentieth Centuries," *Cambridge History of Science*, vol. 5, *The Modern Physical and Mathematical Sciences*, ed. Mary Jo Nye (Cambridge, UK: Cambridge University Press, 2002), 36–53. On the scientific worldview, see David M. Knight, *The Age of Science: The Scientific World View in the Nineteenth Century* (New York: Basil Blackwell, 1986).

6. Ps. 104:19.

7. See the lecture "On the Question: Are the Planets Inhabited?"

8. Halley's Comet returned to within sight of earth in 1835, with predictions, none of them exact, regarding the date of the perihelion, having been made by Marie-Charles-Théodore de Damoiseau, Philippe Gustave Doulcet, Comte de Pontécoulant, Jacob Heinrich Wilhelm Lehmann, Otto August Rosenberger, and Heinrich Wilhelm Olbers. Articles also appeared in the *Mechanics' Magazine*; see, for instance, *Mechanics' Magazine* 23 (April–September 1835): 428. Three years before that there occurred the return of Encke's Comet, which had been identified as a comet in 1819 by Johann Franz Encke. In general, see Stephen Toulmin and June Goodfield, *The Fabric of the Heavens: The Development of Astronomy and Dynamics* (New York: Harper, 1962). Concerning the debates about celestial measurement, see Giuseppe Monaco, *Le comete e l'etere cosmico nell'Ottocento* (Florence: Fondazione Giorgio Ronchi, 2004). For still-useful data, see Amédée Guillemin, *The World of Comets*, trans. James Glaisher (London: S. Low, Marston, Searle, and Rivington, 1877).

9. Boole explored the first of these topics in the treatise that he finished two years after this speech: George Boole, *An Investigation into the Laws of Thought, on Which Are Founded the Mathematical Theories of Logic and Probabilities* (London: Walton and Maberly, 1854).

10. The earliest use of the term "social science" in English is said to have been by an Owenite Irishman, William Thompson, in *An Inquiry into the Principles of the Distribution of Wealth Most Conducive to Human Happiness; Applied to the Newly Proposed System of Voluntary Equality of Wealth* (1824). See Dolores Dooley, *Equality in Community: Sexual Equality in the Writings of William Thompson and Anna Doyle Wheeler* (Cork: Cork University Press, 1996), 361n15.

11. Quoted from memory. The verse from Lucretius, *De rerum natura*, 1:141–142, actually has "quemvis" not "quemcunque." It is translated as "encourage(s) me to face any task however hard. This it is that leads me to stay awake through the quiet of the night" by Ronald Latham in Lucretius, *On the Nature of the Universe* (Baltimore: Penguin Books, 1951), 31.

12. The London cholera epidemic in 1848 inspired physician John Snow to investigate the underlying causes of the spread of the disease in pioneering work carried out between 1849 and 1855. For a useful work, see Steven Johnson, *The Ghost Map: The Story of London's Most Terrifying Epidemic—and How It Changed Science, Cities, and the Modern World* (New York: Riverhead, 2006).

13. Compare Boole, *An Investigation into the Laws of Thought*, 29: "Let it even be granted that the mind is but a succession of states of consciousness, a series of fleeting impressions uncaused from without or from within, emerging out of nothing, and returning into nothing again,—the last refinement of the sceptic intellect,—still, as laws of succession, or at least of a past succession, the results to which observation had led would remain true. They would require to be interpreted into a language from whose vocabulary all such terms as cause and effect, operation and subject, substance and attribute, had been banished; but they would still be valid as scientific truths." Here Boole may have in mind John Stuart Mill, *A System of Logic*, 1st ed. (London: John W. Parker, 1843), book 6, chap. 4.

14. Cicero, *De finibus bonorum et malorum*, book 5.

15. Plato, *Republic*, book 6.

16. Most likely, Virgil, *Georgics*, 2:475–482.

17. Sophocles, *Antigone*, 332–375. Regarding this passage, see Martin Heidegger, "The Ode on Man in Sophocles' Antigone," in *An Introduction to Metaphysics*, trans. Ralph Manheim (New Haven, CT: Yale University Press, 1959), 86–100.

18. The reference is to Aeschylus, *Prometheus Bound*.

19. Again from memory, the citation seems to be drawn from Plato's *Phaedo*, 109D–112A, perhaps in the edition of *Divine Dialogues*, trans. André Dacier, Floyer Sydenham, and Thomas Taylor (London: S. Cornish, 1841), 242–243.

20. See note at the end of the chapter text.

CHAPTER 8

1. Originally published as George Boole, *The Social Aspect of Intellectual Culture: An Address Delivered in the Cork Athenaeum May 29th, 1855 at the Soiree of the Cuvierian Society* (Dublin: George Purcell and Company, 1855).

2. John Milton, *Complete Poems and Major Prose*, ed. Merritt Y. Hughes (New York: Prentice Hall, 1957), 169.

3. George William Frederick Howard, seventh earl of Carlisle (1802–1864), and lord lieutenant of Ireland from 1855 to 1858 and again from 1859 to 1864.

4. That is, the motion of the planets.

5. Edmund Murphy, professor of agriculture at Queen's College and author of *The Agricultural Instructor; or Young Farmer's Classbook, Being an Attempt to Indicate*

the *Connexion of Science with Practice in Agriculture* (Dublin, 1853). Anacharis Alsin-astrum, also known as Elodea canadensis (American or Canadian waterweed), was supposedly introduced into Ireland at the beginning of the century and quickly invaded aquatic areas. Noted in the third edition of Charles Cardale Babington, *Manual of British Botany* (London: Van Voorst, 1851), 304.

6. John Windele (1801–1865) was a local antiquary who published *Historical and Descriptive Notices of the City of Cork and Its Vicinity* (Cork: Bradford and Co., 1839) and was a founding member of the Cuvierian Society.

7. Richard Rolt Brash (1817–1876), a Cork architect and archaeologist, built the Cork Athenaeum. See *Dictionary of Irish Architects 1720–1940*, accessed July 30, 2017, http://www.dia.ie/. Richard Caulfield (1823–1887) was the secretary, librarian, and custodian of the Royal Cork Institution, librarian for Queen's College, and for a time president of the Cuvierian Society.

8. Richard Sainthill (1787–1869) was a wine merchant, antiquarian, and numismatist in Cork, and founding member of the Cuvierian Society.

9. Robert Harkness, FRS and FGS, was a professor of geology at Queen's College from 1853 to 1878.

10. George Ferdinand Shaw (1821–1899) was hired as a professor of natural philosophy at Queen's College after serving as a Trinity College Fellow and dean in Dublin, where he later returned to become the first editor of the *Irish Times* as well as political activist.

11. Francis Jennings was a Cork merchant, member of the Royal Irish Academy and fellow of the Geological Society of London, and among other works, author of *The Present and Future of Ireland as the Cattle Farm of England: and Her Probable Population, with Legislative Remedies* (Dublin: Hodges, Smith, and Co., 1865).

12. Major General Joseph Ellison Portlock (1794–1864) was a British geologist and soldier, author of the *Report on the Geology of the County of Londonderry and of Parts of Tyrone and Fermanagh* (1843) as part of the geologic branch of the Ordnance Survey, and inspector of studies at the Royal Military Academy at Woolwich.

13. Lucretius, *De Rerum Natura*, 2:79. The argument in this paragraph bears comparison to, for instance, William Cooke Taylor, *The Natural History of Society in the Barbarous and Civilized State: An Essay towards Discovering the Origin and Course of Human Improvement* (London: Longman, Orme, Brown, Green, and Longmans, 1840), 1:1–19.

14. The Enlightenment concept of a particular developmental role played by each of the major peoples in the progress of human civilization was expressed notably in Johann Gottfried Herder, *Ideen zur Philosophie der Geschichte der Menschheit* (Riga: Johann Friedrich Hartknoch, 1784–1791), translated by T. Churchill as *Outlines of a Philosophy of the History of Man* (1803), and in Georg

Wilhelm Friedrich Hegel, *Vorlesungen über die Philosophie der Geschichte*, ed. E. Gans (Berlin: Duncker und Humblot, 1837), later translated by John Sebree as *The Philosophy of History* (1857).

15. John Ruskin makes the point in *The Seven Lamps of Architecture* (London: J. Wiley, 1849), 1:192. For modern buildings in the Gothic manner, however, he opts for an English-decorated style modified by French ornamentation.

16. The question was posed by William Whewell in *History of the Inductive Sciences, from the Earliest to the Present Times* (London: J. W. Parker, 1837), 1:78: "The defect was that though they had in their possession facts and ideas, the ideas were not distinct and appropriate to the facts."

17. John Blyth (1815–1892) was a professor of chemistry at Queen's College.

18. Concerning this development, see R. H. Nuttall, "The Achromatic Microscope in the History of Nineteenth-Century Science," *Philosophical Journal* 2 (1974): 71–88.

19. Louis Daguerre introduced his "daguerreotype" process commercially in 1839, and improvements over the next years by Henry Fox Talbot and others increased the popularity of the invention.

20. Boole is evidently here speaking of Jakob Friedrich Fries, *Die Geschichte der Philosophie*, vol. 1 (Halle: Verlag der Buchhandlung des Waisenhauses, 1837).

21. Of the German scholar-diplomat Christian Charles Josias von Bunsen, Boole is certainly referring to the two-volume *Outlines of the Philosophy of Universal History as Applied to Language and Religion*, published in 1854 during the writer's mission to England on behalf of the Prussian king Frederick William IV, drawing on ideas in his own *Ägyptens Stelle in der Weltgeschichte*, published from 1844 and at this time already in its penultimate (fourth) volume (Gotha: F. A. Perthes, 1845).